Common Disorders
of the Hip

Common Disorders of the Hip

Mary C. Singleton and Eleanor F. Branch
Co-Editors

84341

The Haworth Press
New York • London

Common Disorders of the Hip has also been published as *Physical Therapy in Health Care*, Volume 1, Number 1, Fall 1986.

The Haworth Press, Inc., 12 West 32 Street, New York, NY 10001
EUROSPAN/Haworth, 3 Henrietta Street, London WC2E 8LU England

Library of Congress Cataloging-in-Publication Data

Common disorders of the hip.

    "Has also been published as Physical therapy in health care, volume 1, number 1, Fall 1986"—T.p. verso.
    Includes bibliographies.
    1. Hip joint—Diseases—Patients—Rehabilitation. 2. Hip joint—Surgery—Patients—Rehabilitation. 3. Hip joint—Surgery. 4. Physical therapy. I. Singleton, Mary C. II. Branch, Eleanor F.
RD772.C65   1986   617'.376   86-3074
ISBN 0-86656-557-4

Common Disorders of the Hip

Physical Therapy in Health Care
Volume 1, Number 1

CONTENTS

Foreword

The first issue of *Physical Therapy in Health Care* is devoted to the topic of common disorders of the hip. Efforts have been made to provide the reader with an overview of this subject: articles dealing with anatomy, biomechanics, medical evaluation, epidemiology, surgery, and preoperative and postoperative physical therapy are included.

The first article of the issue, on clinical anatomy, lays the foundation for subsequent presentations. Here, basic anatomical features of the hip joint are discussed, including bony structure, capsule, musculature, arterial supply, and innervation. The following paper highlights basic normal biomechanical principles of the hip and relates these to pathological changes which occur in degenerative joint disease and fracture.

The succeeding author elaborates the extent of the problem of hip fracture and its impact on health care resources. He addresses the underlying causes and epidemiology of this disability, emphasizing its challenge to the physical therapist.

The physician author of the ensuing paper discusses evaluation and management of pain—a common symptom of hip disease. He describes particular methods of assessment, etiologic considerations, and factors which influence the choice of treatment.

A surgeon delineates the changing picture of surgical intervention for arthritis of the hip through illustrative clinical examples. He describes briefly nonoperative management and follows with a description of various surgical procedures and their implications for the physical therapist.

The reported results of a survey of physical therapy facilities in Florida, determining the extent of use of preoperative evaluation procedures and instructions for patients receiving hip joint replacement, constitute the next article. The authors suggest that further studies are required to document the objective benefits of preoperative physical therapy intervention. The authors of the seventh article in the issue present physical therapy guidelines for management of patients with hip fracture and joint replacement.

They discuss the physical therapist's role in screening for hip fracture and risk factors for non-recovery from this condition. Included are rehabilitation programs for patients following hip arthroplasty, as well as postoperative complications of which the physical therapist should be aware.

To conclude this issue, a physical therapist who has undergone a hip joint replacement shares with the reader her personal experience. This article portrays the author's perceptions of some of the attitudes of health professionals toward patients who have undergone such surgery.

Preface

Physical Therapy in Health Care, a new quarterly journal, is intended as a vehicle for the dissemination of information on the comprehensive management of types of disability frequently encountered in the everyday practice of physical therapy. It is not designed to serve as a journal for the publication of pure research in this field. Each issue will be devoted to a selected theme topic, thus providing a basic informational resource on the particular subject. Although the major thrust of the journal will be on evaluation and treatment, related anatomical, physiological, and pathological aspects of the topic also will be addressed. Although this periodical is intended primarily for the physical therapist reader, the editors anticipate that other health professionals will find it a valuable addition to their libraries and will be encouraged to submit articles for possible publication.

The editors believe that *Physical Therapy in Health Care* can be made more meaningful to the general readership if health professionals will submit their suggestions for topics to be themes of future issues. Our hope is that it will help to meet the need, expressed by numbers of physical therapists, for an appropriate avenue through which they may contribute to the physical therapy literature.

Common Disorders of the Hip

The Hip Joint:
Clinical Oriented Anatomy—A Review

Mary C. Singleton, Ph.D., P. T.

ABSTRACT. This article discusses the basic anatomical features of the hip joint. Structures providing stability to the joint are articulating surfaces of the bones, the joint capsule and its ligaments. The three major bursae around the hip joint consisting of the trochanteric, iliopectineal, and the ischiogluteal bursae are described and clinical effects of bursitis cited. Muscles and major movements of the hip are summarized. The pattern of arterial distribution to the upper end of the femur is outlined and effects of its interruption discussed. The major nerves to the hip joint are described as branches from the femoral, obturator, accessory obturator, superior gluteal nerves and the nerve to the quadratus femoris muscle.

In any consideration of the hip joint, a thorough understanding of the basic anatomical features of the joint is necessary before functional and clinical aspects can be assessed. This article presents a review of outstanding structural characteristics of the hip joint and a discussion of those features which affect normal function or develop disability.

Since the major functions of the hip joint are to provide stability for support of the trunk and weight bearing for body movement, the structural elements basic to these functions are of prime importance.

JOINT STRUCTURES PROVIDING STABILITY

Articulating Surfaces of Bone

The mechanical arrangement of bones of the hip insures stability while not limiting freedom of movement. The hip joint is inherently

---

Dr. Singleton is Professor Emeritus, Division of Physical Therapy, University of North Carolina, Chapel Hill, NC 27514.

strong, the structure being well adapted to supporting the weight of the trunk as well as permitting a functional degree of mobility.[1] The joint is formed by the head of the femur fitting snugly into the cup-shaped acetabulum.

The head of the femur forms more than half a sphere. It is covered by articular cartilage except over a small roughened pit, the fovea of the femoral head, where the ligament of the head of the femur is attached (Figure 1).[1,2] The head is fitted closely into the acetabulum for an area extending over nearly half a sphere to form a typical ball and socket synovial joint.[1]

All three elements of the hip bone (ilium, ischium, and pubis) contribute to formation of the acetabulum.[2] The inner surface of the acetabulum is not lined completely with articular cartilage as the portion of the surface covered by articular cartilage forms an incomplete ring termed the lunate surface. This cartilaginous ring is broadest and thickest at its upper part where pressure of the body weight falls when the body is in an erect position and narrowest at its lower portion where it covers

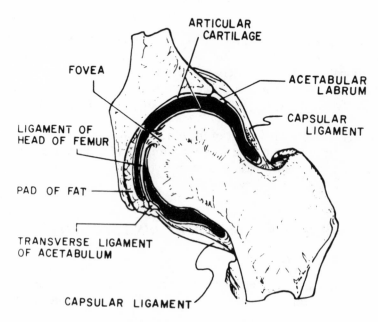

FIGURE 1. A section through the hip joint.

the pubic constituent. The floor of the acetabular fossa within the lunate ring is devoid of articular cartilage. This area is occupied by a mass of fatty tissue covered by synovial membrane (Figure 2).[1,2] Such a structural arrangement of specific areas of weight bearing articular cartilage in the acetabulum becomes important to consider when analysis of particular weight bearing areas of the hip joint is attempted.

At the margin of the bony acetabular cup, the head is closely embraced by the acetabular labrum which contracts the orifice and deepens the surfaces of articulation (Figure 2). This fibrocartilaginous structure holds the femoral head in place even when the surrounding capsular fibers have been divided.[1] The acetabulum is deficient inferiorly, having a deep indentation, the acetabular notch. This notch is converted into a foramen by the transverse acetabular ligament and through the foramen pass nutrient vessels and nerves into the joint.[1]

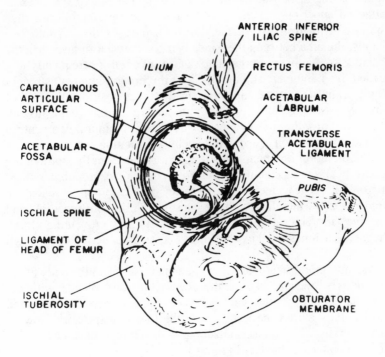

FIGURE 2. The acetabulum and surrounding area.

Joint Capsule and Ligaments

Capsule. The fibrous capsule of the hip joint is rough, strong and dense, thickest at the upper forepart of the joint where the greatest amount of resistance is required for the upright position. Sets of fibers are arranged in circular and longitudinal orientation. The circular fibers are deeper and form a collar around the neck of the femur. The longitudinal fibers are greatest in number at the upper part of the capsule where they are reinforced by the iliofemoral ligament.[1,2] From their attachment to the front of the femoral neck, many of the longitudinal fibers are reflected upward along the neck as longitudinal bands called retinacula. These contain blood vessels supplying the head and neck of the femur.[1]

Particular areas of the fibrous capsule are reinforced by ligaments which blend intimately with the capsular fibers.

Iliofemoral ligament. This ligament, which lies anterior to the hip joint, is the strongest of all ligaments in the body.[2] Its apex is attached superiorly to the lower part of the anterior inferior iliac spine and to the body of the ilium. Distally, it divides into two bands, the area between the bands being somewhat thinner than the bands. As it appears in the shape of an inverted Y, frequently it is called the Y ligament (Figure 3).[1] The iliofemoral ligament strongly limits extension at the hip joint. In the erect position, the line of gravity passes behind the hip joint and the trunk tends to fall backward or rather to rotate backward. The iliofemoral ligament, in its anterior position, checks the backward rotation.[1,3]

Pubofemoral ligament. This ligament arises from the body of the pubis near the acetabulum and from the adjacent pubic ramus. Below, it passes anterior to the head of the femur to reach the femoral neck, blending with the capsule and the more medial band of the iliofemoral ligament. It assists in preventing trunk hyperextension as well as checking excessive abduction of the thigh (Figure 3).[1]

Ischiofemoral ligament. This ligament arises from the body of the ischium below and behind the acetabulum and blends with the circular fibers of the capsule. The upper fibers are oriented horizontally across the joint, while the lower fibers spiral upward and laterally with both sets attaching to the femoral neck just medial to the greater trochanter (Figure 4).[1]

The ligament of the head of the femur. One may be surprised upon study or dissection of the hip joint to find that the ligament of

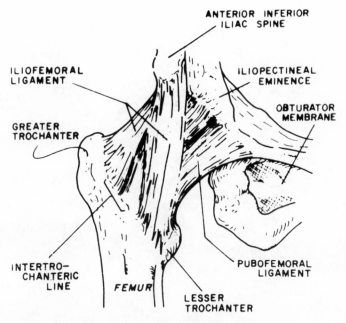

FIGURE 3. The capsule of the right hip—anterior aspect.

the head of the femur seems to lack the strength to demonstrate any definite ligamentous function. It consists of weak connective tissue fiber surrounded by synovial membrane which is attached to both sides of the acetabular notch and to the floor of the acetabular fossa deep to the transverse ligament. From this origin, the ligament of the head of the femur passes around the articular surface of the head to insert into the upper part of the fovea (Figure 1). Other than being the structure which transmits the artery of the ligament of the head to the fovea, the function of this ligament is uncertain (Figure 7). No appreciable disability appears to result from its rupture or absence.[4,5] Aseptic necrosis, however, may occur in a small portion of the femoral head following a rupture of this ligament which entails cutting off the blood supply usually afforded the region of the fovea by the vessels in the ligament.[5]

Although quite strong, the capsule of the hip joint shows a weak area anteriorly in an intermediate region between the iliofemoral and the pubofemoral ligaments. Not a ligament, but guarding this critical area, is the tendon of the psoas (iliopsoas) muscle (Figure 5). This

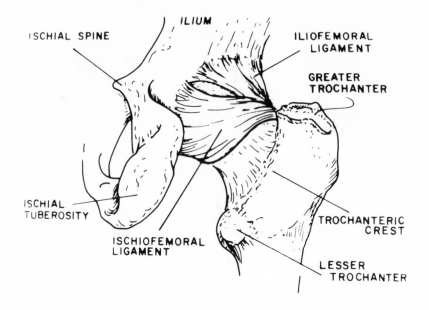

FIGURE 4. The capsule of the right hip joint—posterior aspect.

tendon provides important support at a location otherwise deficient in ligamentous protection. It assists the iliofemoral ligament as an anterior force resisting extension of the trunk at the hip joint.[6,3]

BURSAE

Closely related to the joint capsule are muscle attachments and their bursae. The latter are closed sacs lined by specialized connective tissue containing synovial fluid. Bursae are usually found over bony prominences, especially where a muscle or tendon moves over a projection of bone.[7] The function of bursae is to facilitate gliding movements by diminishing friction. As many as eighteen bursae around the hip have been described but only three are of major importance. These are the trochanteric bursa, the iliopectineal bursa, and the ischiogluteal bursa (Figure 6).[8]

Trochanteric bursa. This is a large bursa located between the tendon of insertion of the gluteus maximus and the posterolateral prominence of the greater trochanter. An indication of bursitis is tenderness over the bursa elicited by pressure or by tensing the muscle when the hip is flexed and internally rotated.

Iliopectineal bursa. This is the largest and most constant bursa about the hip. It lies between the iliopsoas muscle anteriorly and the iliopectineal eminence posteriorly. It is situated lateral to the femoral vessels and overlies the capsule of the hip. Frequently, the bursa communicates with the joint cavity and is involved simultaneously with the synovium.[8] When the bursa is inflamed, iliopsoas contraction, as in active hip flexion or in a stretching of this muscle, as in hip extension, elicits pain in the region.

Ischiogluteal bursa. This bursa overlies the ischial tuberosity and becomes inflamed in persons whose occupation demands prolonged sitting (Tailors or Weavers Bottom). Irritation of the adjacent sciatic nerve may produce symptoms of sciatica. Avoidance of pressure generally relieves the symptoms.[8]

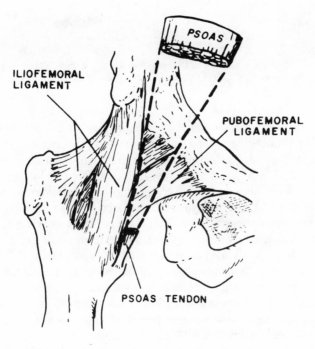

FIGURE 5. The hip joint capsule—anterior view showing the tendon of the psoas muscle guarding the weak point.

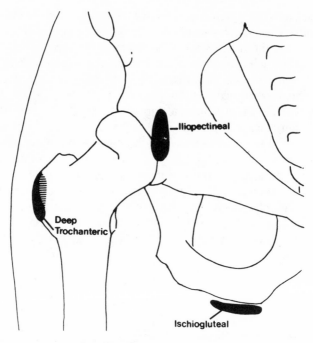

FIGURE 6. Most commonly affected bursa about the hip. (Adapted from Raney, RB; Brasher, HR; Shands Handbook of Orthopedic Surgery, ed. 8, 1971, p. 350.)

Bursitis. Because the three primary bursae of the hip joint are related physiologically and developmentally to tendon sheaths, they, therefore, are subject to similar disorders (traumatic inflammation, infection, etc.). The usual bursitis is an inflammatory reaction to overuse or excessive pressure and subsides with rest, moist heat, or needle puncture.[8] Close anatomical relationships of the iliopectineal bursa with the femoral nerve can result in a femoral neuropathy caused by pressure on the nerve from cystic enlargement of this bursa. Such a case is reported by Lavyne et al.[9] These authors also note that any openings of the hip joint capsule allow the bursa to be continuous with the synovial membrane of the joint. This condition permits accumulation of joint fluid to extend into the iliopectineal bursa so that it can compress adjacent structures.

An editorial in *Lancet*, 1983,[10] states that although trochanteric bursitis is common, little published work is available on the subject. It quotes Sweney[11] who says that bursitis at this location probably is related to altered hip joint mechanics. Sweney also suggests that such bursitis may occur in patients who have had total hip replacements where prosthetic dysfunction may be suspected thus obscuring the true explanation of a treatable trochanteric bursitis.

MUSCLES AND MOVEMENT

Muscles around the hip joint, in addition to being responsible for its movement and strength, also may play a postural role, assisting the ligaments in providing stability.[6] All movements typical of a ball and socket multiaxial joint are possible—flexion, extension, abduction, adduction, lateral (external) rotation, and medial (internal) rotation, and circumduction. When the femur is in slight flexion at the hip joint, the pelvis is prevented from rolling downward on the femoral head by the action of the posterior hamstring muscles. A number of investigators have shown that, in the position of easy standing, the gluteus maximus does not function as a postural muscle. This muscle is active more particularly when the thigh is extended against resistance as in rising from a bending or sitting position or when walking up steps.[2,3]

All three of the major ligaments of the hip capsule become tight on trunk or femoral extension (as in the standing position). They become tighter as the joint approaches its position of "close-packing." The position of hip extension "winds up" the spirally running fibers of the capsule and thereby forces the head of the femur deeper into the socket and is self-arresting. So strong are the iliofemoral and the ischiofemoral ligaments that forcible attempts to produce hyperextension of the joint accompanied by forward pressure on the iliac crest result in movement at the sacroiliac joint.[2]

Abduction of the femur is a relatively free movement limited by tension on the muscles which adduct the femur, by the pubofemoral ligament, and by the medial band of the iliofemoral ligament. The function of the abductor muscles is to prevent the pelvis from becoming adducted—that is to prevent the body from falling to the unsupported side when one foot is off the ground.[3]

Adduction of the femur is limited by contact with the opposite limb but a wider range is possible when the thigh is flexed. Adduction of the flexed thigh is limited by tension of the muscles which abduct the femur and by the lateral band of the iliofemoral ligament. Medial rotation of the femur is limited by the muscles which laterally rotate the femur, by the iliofemoral ligament and by the posterior part of the fibrous capsule. Lateral rotation of the femur is a powerful movement. It is limited by tension on the muscles which medially rotate the femur and by the lateral band of the iliofemoral ligament.[2]

In contrast to the strong position of extension at the hip joint, the flexed position is relatively weak. This is true because of the relative weaker portion of the posterior portion of the joint capsule and that, when the femur is flexed at the hip joint, all parts of the fibrous capsule are released. Because of this laxity of the capsule and ligaments, dislocation of the femur posteriorly is likely to occur if enough force in that direction is put upon the distal end of the flexed femur.[4] Such a condition is commonly produced by a hard blow upon the knee of a sitting individual.[5] If the hip is adducted simultaneously as such a force is applied, the head of the femur is unsupported posteriorly by the acetabulum and dislocation can occur without an associated acetabular fracture. If the hip is abducted, dislocation must be accompanied by a fracture of the posterior acetabular lip.[12]

BLOOD SUPPLY

Practical relevance of the pattern of arterial distribution to pathology of the upper end of the femur has stimulated considerable investigation of the precise arrangement of these vessels. Many such studies have been done in relation to the results of fracture of the femoral neck which interrupts the arteries of that region and thus deprives the femoral head of an adequate blood supply.[6]

Arteries of the hip joint are derived from: the medial femoral circumflex branch of the profunda femoral artery; the obturator artery by way of the acetabular branch which sends a small vessel along the ligament of the head of the femur; and the superior and inferior gluteal arteries.[1] The general pattern of blood supply to the joint is shown in Figure 7.

Harty describes the basic pattern of anastomoses found at the upper end of the femur as three major anastomotic circles surrounding the hip joint:

1. On the acetabular margin with its blood supply provided by the obturator, superior and inferior gluteal, and the medial

femoral circumflex arteries. It supplies the adjacent muscles, capsule, and underlying bone;

2. In the subcapital sulcus, floating in a fat annulus and covered by synovial membrane. This anastomosis receives its blood supply from the retinacular vessels and is distributed to the femoral head and underlying neck;

3. An anastomotic ring at the base of the neck which contributes to major vessels designated for the neck and femoral head.[14]

This author states that the largest and most constant vessel of supply

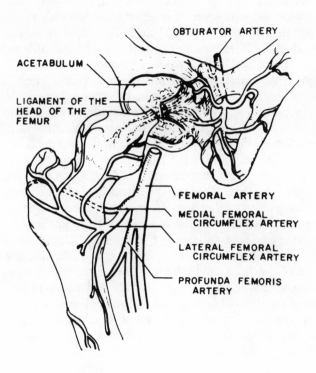

FIGURE 7. Major blood supply to the upper end of the femur. (Schematic drawing adapted from Henderson, MS; Surg Clin North Am, Aug. 1939, p. 927.)

to the region is the medial circumflex artery which passes dorsally between the pectineus and the iliopsoas muscles, then between the tendons of the obturator externus and the iliopsoas.

As stated previously, reflections of the joint capsule form bands which contain vessels passing into the joint. The retinacular arteries have been described by Sevitt, according to their location, as superior and inferior groups. These arteries, which are derived from the circumflex femoral arteries, especially the medial one, pass medially to the head under the synovium of the neck. Superior retinacular arteries and their branches are found by this investigator to be the most important arterial supply to the head of the femur.[15]

The importance of the artery of the ligament of the head of the femur has been evaluated in various ways. Sevitt's study showed that the artery in this ligament was either absent or unimportant for the head in most subjects (at least in elderly adults). Either the vessels in the ligament never reached the head or they supplied only a limited subfoveal zone.[15] Trueta and Harrison showed that although ligamentous vessels reached the head of the femur after birth, they did not join in its circulation for three to four years. These authors concluded that the vessels contributed to the capital circulation in adult life.[16] Harty says that the upper femur derives its main blood supply from the subcapital anastomosis inasmuch as the contribution via the artery of the ligament of the head, at any age, is limited to a small medial segment of the head of the femur.[13] Clemente, in *Gray's Anatomy*, states that the artery of the femoral head is of great importance until puberty but, as age advances and anastomotic channels develop among other vessels, this artery becomes smaller and less significant to the vascular integrity of the joint.[1]

General agreement appears to exist among investigators regarding the uncertain presence, site, and extent of the arteries of the ligament of the head of the femur. Therefore, interruption of the major arteries of the femoral neck through dislocation of the joint or fracture of the femoral neck usually results in loss of viability of the femoral head as it becomes deprived of its adequate blood supply. This is especially true in elderly patients, those most likely to sustain such fractures.

INNERVATION

Since intense pain, often associated with hip pathology, prevents normal weight bearing, the pattern of hip joint innervation becomes an important factor in function. The nerves to the hip joint come from the femoral, obturator, accessory obturator (when present), nerve to the quadratus femoris, and the superior gluteal nerve.[1]

Gardner described common types of individual variations of innervation:

1. Branches from the femoral nerve to the iliofemoral ligament, to the posterior part of the capsule, and to the region of the pubofemoral ligament;
2. A distribution from the obturator and accessory obturator nerves to the region of the pubofemoral ligament:
3. A branch of the superior gluteal nerve to the superior lateral region of the capsule;
4. Branches of the nerve to the quadratus femoris supplying the posterior capsule including the ischiofemoral ligament (Figure 8). Gardner reported that less overlap of innervation tends to occur at the hip joint than is found in some other joints and that many of the hip branches seem to be vascular ones.[17]

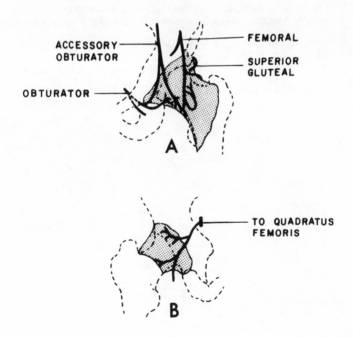

FIGURE 8. Anterior (A) and posterior (B) views of types of nerves to the hip joint. (Schematic drawing adapted from Gardner, E; Anat Rec 101: 353,1948.)

REFERENCES

1. Clemente CD: Anatomy of the Human Body by Henry Gray, ed 30 (American). Philadelphia, PA, Lea and Febiger, 1985, pp 390–393; 395

2. Williams PL, Warwick R: Gray's Anatomy, ed 36. Philadelphia, PA, WB Saunders Co, 1980, pp 383; 477–481

3. Basmajian JF: Grant's Method of Anatomy, ed 10. Baltimore, MD, The Williams and Wilkins Co, 1980, pp 293–296

4. Romaines GJ: Cunningham's Textbook of Anatomy. London, Oxford University Press, Inc, 1972, pp 172; 240–241; 663

5. Hollinshead WH: Anatomy for Surgeons, ed 2. New York, Hoeber Medical Division Harper and Row, 1969, pp 663–678

6. Singleton MC, LeVeau B: The Hip Joint—Structure, Stability and Stress, A review. Phys Ther 55:957–973, 1975

7. Shands AR, Raney RB: Handbook of Orthopedic Surgery. CV Mosby Co, St. Louis, MO, 1937, p 24

8. Turek SL: Orthopedics—Principles and Their Application, ed 4. JB Lippincott Co, 1984, pp 1112; 1245–1246

9. Lavyne MH, Voorhies RM, Call RH: Femoral Neuropathy caused by an iliopsoas bursal cyst. J. Neurosurg 56: 584–586, 1982

10. Editorial: Lancet, Feb 5, 1983, p 281

11. Sweney RL: Pseudoradiculopathy in subacute trochanteric bursitis of the subgluteus maximus bursa. Arch Phys Med Rehab, 57: 387–90, 1976

12. Ellis H: Clinical Anatomy, Blackwell Scientific Publications. Oxford, England, 1983, p 274

13. Harty M: Blood supply of the femoral head. Br Med J, II; 1236, 1953

14. Harty M: Anatomic considerations. Ortho Clinics of North Am, Vol 1, #4, Oct. 1982

15. Sevitt S: The distribution and anastomosis of arteries supplying the head and neck of the femur. J Bone Jt Surg, 47B: 560, 1965

16. Trueta J, Harrison MHM: The normal vascular anatomy of the femoral head in adult man. J Bone Jt Surg (Br), 35: 442–461, 1953

17. Gardner E: The innervation of the hip joint. Anat Rec, 101: 353–371, 1948

Basic Biomechanical Principles of the Hip and Their Relationship to Degenerative Joint Disease and Fracture: A Review

Barney F. LeVeau, Ph.D., P.T.

ABSTRACT. The purpose of this paper is to present the normal function and basic biomechanical principles of the hip and to relate these to the pathological changes which occur in degenerative joint disease and fracture. Stresses and strains upon the hip joint are discussed in relation to the structural elements of the hip, the magnitude and frequency of the applied load, and the way in which the load is applied. The relevance of mechanical principles to cause and prevention of degenerative joint disease and fracture are presented.

Biomechanical principles play an important role in the prevention and treatment of disorders of the hip. Degenerative joint disease (DJD) and fractures are major problems associated with the hip joint. The consequences of these disturbances, however, may be reduced or eliminated if correct body mechanics are employed. Biomechanics and pathological processes are closely related, in that poor mechanical use can precipitate or intensify pathological changes, while pathological alterations can contribute to poor mechanical function. This paper discusses normal function and the basic biomechanical principles of the hip and relates them to pathological changes which occur in DJD and fracture.

The function of the hip joint is to provide a stable center for motion between the trunk and the thigh. This function allows an

Dr. LeVeau is a professor and Chairman of the Department of Physical Therapy in the School of Allied Health Sciences, University of Texas Health Science Center at Dallas—Dallas, TX 75235.

individual to ambulate from place to place, to sit and stand, and to participate in a wide range of occupational and recreational activities. Since the configuration of the joint is a ball and socket, hip motion occurs in all planes. The ranges of motion, however, may be influenced by abnormal bone or soft tissue in the region.[1] Abnormalities of pelvic width, acetabular size, shape, and obliquity, and the angle of inclination and anteversion of the proximal end of the femur all may influence motion. The major soft tissues affecting the amount of motion occurring at the hip are the muscles and ligaments spanning the joint.

GENERAL BIOMECHANICAL PRINCIPLES

Gravity, muscle action, and inertia are the forces which produce or limit motion at the hip joint. These forces control motion at the joint but also produce loads of great magnitude, setting up stresses within the joint. These stresses in general are the body part's reaction to loads applied to it. Stress, which is the force per unit area, can be compression, tension, or shearing. In compression, the forces act along the same line and toward each other. In tension, the forces are along the same line, but directed away from each other. In shearing, the forces do not act along the same line, but are parallel and tend to slide past each other. All of these stresses may occur at the same time within the hip (Figure 1).

Strain is the deformation of a body part which results from a load being placed on it. The magnitude of strain depends upon the load applied and upon the elasticity of the tissue. Within certain limits of loading, the tissue will deform (strain) in direct proportion to the load applied and to the resulting stress within the tissue. Stiffness is the property of a material that resists deformation when the material is loaded. In contrast, compliance of a material is the readiness with which it strains as a load is applied to it (Figure 2).

The stresses and strains developed within the articular cartilage and bone of the hip joint depend upon (1) the structural elements of the hip, (2) the magnitude and frequency of the applied load, and (3) the way in which the load is applied.[2]

Structural Elements of the Hip

The head of the femur is covered with a thin layer of articular cartilage, while the cartilage of the acetabulum is horseshoe-shaped,

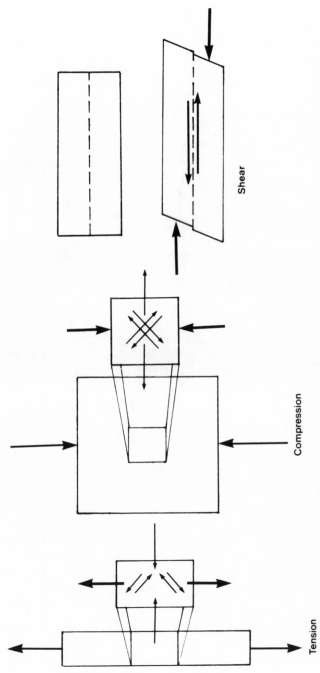

Tension   **Compression**   **Shear**

FIGURE 1. The principal stresses and strains of tension (pulling apart), compression (pushing together), and shear (sliding past).

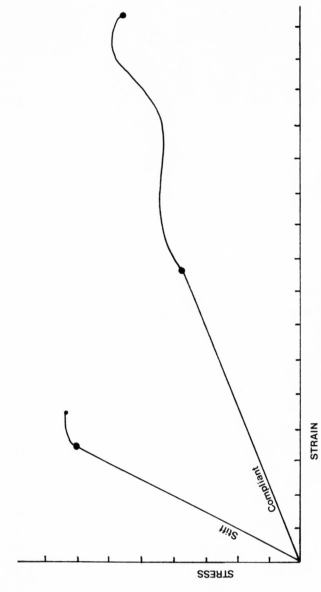

STRAIN

FIGURE 2. Stress-strain curves showing examples of stiff and compliant materials.

incompletely covering its surface. Articular (hyaline) cartilage is composed of 60–80% water; most of the remainder of this tissue is split into a 60 to 40 ratio between collagen fibers and proteoglycan complexes. The deepest layer of articular cartilage is calcified and abuts on subchondral bone (Figure 3). The subchondral bone is supported from below by trabecular bone which in turn is held firm by more distant cortical bone. The surface between the cartilage and subchondral bone is crooked and uneven. This irregular interface is necessary to reduce the shear strain which would occur if the interface were a smooth line.[2,3,4] At the bone-cartilage interface, the collagen fibers are perpendicular to the interface, providing support for the compressive loads applied to the joint. Nearer the joint surface, however, fibers lie parallel to the surface, to allow for the shearing stress which occurs in that location (Figure 3). This surface acts as a wear-resistant layer for the joint.[3,4,5]

One of the functions of articular cartilage is protection of the underlying bone from high stress. Normal cartilage deforms about 20 times more easily than cancellous bone, providing an increase in the contact area for loading[4,5] and reducing the compression stress within the articular cartilage and the underlying bone. Without cartilage, joint loading would occur within a relatively small area. The even distribution of the load by the presence of cartilage prevents the development of localized high compression areas which might otherwise occur because of the irregular interface between cartilage and subchondral bone.

In addition, the system of tiny bony trabeculae below the cartilage provides an effective mechanical support for the stress transmitted to bone via the cartilage. The area of subchondral bone and connecting trabeculae is greater than that of the supporting cortical bone; because of this increased area, the compressive stress in the end of the bone is reduced. Trabeculae also are less stiff than cortical bone and allow for greater energy storage during dynamic loading. Thus, the overall joint reaction to a load progresses from minimal stiffness in the articular cartilage to maximum stiffness in the cortical bone.[3,5]

Cartilage possesses both viscous and elastic properties. The tissue tends to resist the flow of water through it, but the reaction to loading is time-dependent. During a slowly applied load, fluid in the cartilage will gradually flow away from the loaded area and, as the load is released, the cartilage will regain its original shape with the return of the fluid. Rapid loading of the cartilage, however, will not

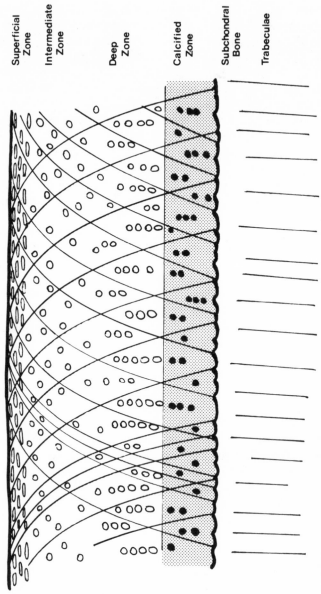

FIGURE 3. Schematic drawing of the layers of cartilage from the superficial zone to its interface with subchondral bone and trabecular support. Note the uneven surface between the calcified zone of cartilage and the subchondral bone. Also note that the cartilage fibers change direction from columnar, or perpendicular to the surface, in the deep zone, to parallel to the surface in the superficial zone.

Superficial
Zone

Intermediate
Zone

Deep
Zone

Calcified
Zone

Subchondral
Bone

Trabeculae

20

allow time for this to occur and the tissue will respond in a purely elastic fashion.[4,6] The viscoelastic nature of cartilage, in addition to its relative stiffness and thinness, accounts for its inability to absorb rapid dynamic loads. Bone attenuates (absorbs) dynamic forces more effectively than cartilage. As stated previously, during dynamic loading bony trabeculae allow for greater energy storage than does stiffer cortical bone. This energy storage capacity, however, varies with the speed of loading. The faster the bone is loaded, the less compliance it demonstrates and the less energy can be stored; microfractures of trabeculae may result.[7]

Results of such time-dependent reactions of the tissues can be demonstrated by the practical example of the speed of gait. Walking about one step per second is too rapid for the viscoelastic property of cartilage to be effective. Hence, during such dynamic activities, the behavior of cartilage is purely elastic and of little value for absorbing large loads, which are therefore absorbed by underlying bone. Since cartilage provides a large area over which bone is loaded, however, the stresses within the bone are of lower magnitude than if cartilage were not present.

As cartilage does not absorb energy effectively in dynamic activities, and too much stress within bone can lead to microfractures, another system must be used to protect the joint. This system involves muscle contraction. For example, slight knee flexion at heel strike in normal gait helps to protect the hip joint from excessive stress because the active contraction of muscles crossing the hip and knee can absorb large amounts of energy. If the knee is not allowed to flex as the dynamic load is applied, the shock-absorbing mechanism of the contracting muscles is unavailable and the subchondral bone will respond with greater stress.[6,8]

During gait, the compression load upon the hip joint can reach a magnitude of 4 to 7 times body weight in a time less than 0.5 seconds (impulse loading). In the normal hip, however, the compression stress of cancellous bone in the subchondral area of the joint is near its ultimate stress level with a load of three times body weight;[4,9] injury to the hip is minimized by the structural elements discussed.

Magnitude and Frequency of Loading

If only a small load is applied to the hip joint and is released, the surrounding tissues will deform (strain) and then return to their original size and shape (elasticity). If the load exceeds the tissue's ability to withstand the load (stress), permanent deformation or breakdown of the tissue will occur. If the load is applied and released rapidly, as during running or walking, energy is lost into the tissues in the form of heat. The rapid cyclic loading and unloading of the tissue can cause the failure of the tissue from mechanical fatigue. The summation of several rapid loads, although they may be under the elastic limit, can result in a fatigue breakdown of a part of the tissue; this means that greater stress is produced in the remainder of the tissue, which may subsequently fail.[6]

The Way Loads Are Applied

Stresses are established within the hip joint and the proximal end of the femur by direct loading and by bending moments. Direct loading of the joints causes compression perpendicular to the area of joint contact and shear parallel to the joint surface. Tension, as well as increased shearing, is set up within the cartilage as movement of the loaded joint occurs.[6] The proximal end of the femur resembles a cantilever, or eccentrically loaded beam. A load on the head of the femur caused by body weight and muscle forces creates a bending moment within the neck of the femur.[10] The intersection of the anatomical axis and the axis of the femoral neck serves as the axis of this moment. The magnitude of the bending moment depends upon the magnitude (F) of the applied load and the length (d) of the lever arm (Fxd) (Figure 4). The bending moment is opposed by a resisting moment which is related to the stresses developed within the femoral neck. Tension develops within the superior regions of the neck, while compression is located within the inferior aspect of the neck (Figure 5). The magnitude of the tension and compression stresses along the femoral neck depends upon the magnitude of the bending moment and the cross sectional area of the femoral neck. The stress at any point along the neck depends upon its distance from the femoral neck axis. The stress is greatest in the subtrochanteric region.[2]

The formation of cortical bone and trabeculae is a reflection of the body's reaction to stress. The greater the load applied, the greater the stress is needed to withstand the load. The body reacts by increasing the amount of bony material in the area. Denser trabeculae and cortical bone in the joint area indicate that greater

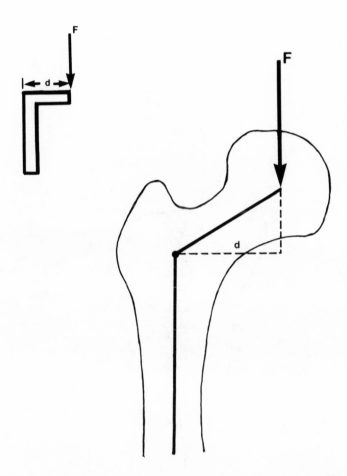

FIGURE 4. A load (F) applied to the head of the femur caused by body weight and muscle force creating a bending moment (Fxd) within the neck of the femur.

loads are being carried by these areas. The increase in density in the hip region is usually in response to compressive loads but also may be a response to tensile loads (Figure 6).

The most common example of muscle contraction and body weight influencing the load across the hip joint is that of single-leg stance (Figure 7).[1,5,12–15] The combined weight of the superincumbent parts establishes a moment of force around the supporting hip joint. This moment must be resisted by a moment created by muscle

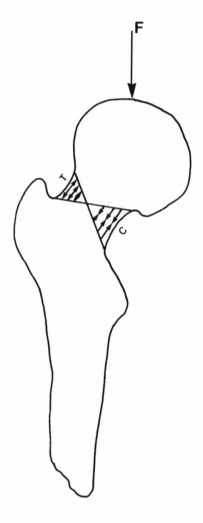

FIGURE 5. As a load (F) is applied to the head of the femur, tension (T) develops within the superior aspect and compression (C) develops within the inferior aspect of the femoral neck.

force. The magnitude of the weight acting across the single supporting hip joint is more than doubled from the situation of bilateral stance; also, a relatively large amount of muscle force adds to this load. The magnitude of the muscle force (M) depends upon the superincumbent weight (W), the lever arm length to a line of

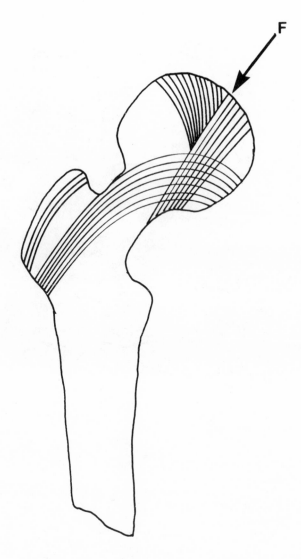

FIGURE 6. Trabecular patterns within the proximal end of the femur set up by tension and compression stresses.

force of this weight (Lw), and the lever arm length (Lm) to the muscle. Moving the body parts closer to the supporting hip joint will decrease the lever arm length of the superincumbent weight and

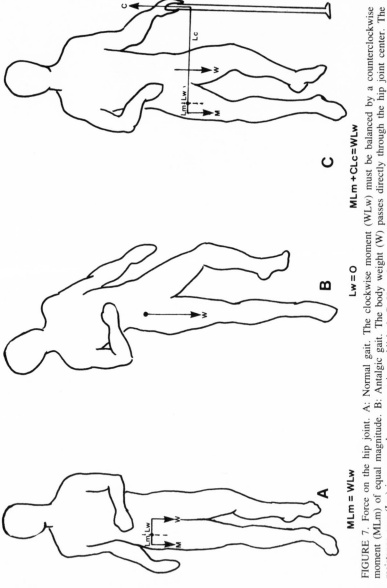

MLm = WLw   Lw = O   MLm + CLc = WLw

FIGURE 7. Force on the hip joint. A: Normal gait. The clockwise moment (MLm) of equal magnitude. B: Antalgic gait. The body weight (W) passes directly through the hip joint center. The resistance arm (Lw) is zero and no moment is established. C: Use of a cane. The clockwise moment (WLw) is balanced by the moments from the abductor muscles (MLm) and the cane (CLc).

26

decrease the magnitude of muscle force needed. An example of this is the antalgic gait of a patient with a painful hip.[15] Another way to reduce the load on the hip joint is by having the patient use a cane in the hand opposite the painful hip. The moment established by the force on the cane and its relatively long lever arm from the hip joint will counteract much of the effect of the body weight. Therefore, a smaller amount of abductor muscle force is needed to maintain the position of the pelvis and less force is borne by the hip joint (Figure 7).[13]

Rarely are other examples of muscle force acting across the hip joint evident. Any time, however, that the line of force of the superincumbent weight falls outside the joint axis in single-leg support, or anterior or posterior to the joint in either single-leg or double-leg support, muscle force is needed to counteract the moment. This muscle force will increase the load upon the joint. Any time a muscle crossing the hip joint contracts, the load on the joint will be affected. The magnitude of the load depends upon the mechanical aspects mentioned earlier. These include the leverage configuration, the weight of the body part involved, and the magnitude of the muscle force. Examples include forward bending from the erect position and straight leg raising from the supine position (Figures 8, 9).

DEGENERATIVE JOINT DISEASE AND FRACTURE

In the disorders of degenerative joint disease and fracture, basic pathological alterations of the normal tissues result in distortion of normal biomechanical functions of the hip.

Degenerative Joint Disease

Degenerative joint disease (DJD) basically can be described as a progressive deterioration of articular cartilage. During the degenerative process, cartilage becomes fibrillated and the cartilage fibers torn, leading to softer cartilage. The major manifestations of the disease include degeneration of the articular cartilage, thickening of subchondral bone, remodeling of marginal and central subchondral bone, and formation of spurs and subarticular bone cysts.[16,17] Osteoarthritis and osteoarthrosis are synonymous with the term degenerative joint disease, but DJD describes more precisely the pathological changes occurring.[18,19]

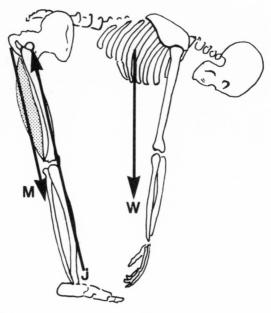

FIGURE 8. In forward bending from the erect position, the downward forces of the upper body combined weight (W) and the hip extensor muscles (M) apply a load to the hip joint (J).

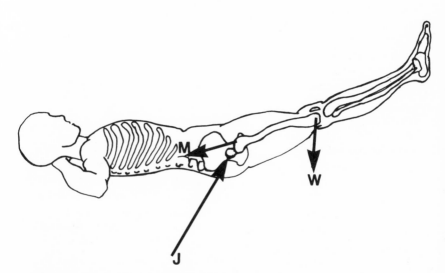

FIGURE 9. During active single leg raising, the combined forces of the limb weight (W) and the hip flexor muscles (M) apply a load to the hip joint (J).

Degenerative joint disease is characterized by gradual development of joint pain and limited range of motion.[18] Limitation of ranges of motion of the hip occurs in extension, internal rotation, and extreme flexion.[20] Some hip joints also may be limited by adduction and external rotation contractures.[19]

The causes of DJD are unclear. One mechanism may be related to an inherent weakness of the articular cartilage; another may be related to abnormal mechanical stresses set up within the joint.[21] Probably both are involved. Authors have separated the etiology of DJD into two major divisions, primary and secondary.[17-19,20,22] The primary type occurs without any known preexisting abnormalities or causative factors; it implies idiopathic degeneration. In the secondary type, major underlying causative factors are joint degeneration secondary to a known injury, deformity, or disease.[19,22] No matter what factors initiate the lesion, mechanical factors are involved from early stages of the disorder.[23] Although the disorder usually becomes manifest in later years, the causes of DJD may be present before birth. Many types of deformity, injury, and disease in childhood can produce the initial cartilaginous lesion leading to DJD in the adult. The most common cause of secondary DJD is deformity or subluxation of the hip, including acetabular dysplasia and femoral head tilt.[20,21] Another important structural factor may be flattening of the femoral head and progressive subluxation of the hip.[22] Recent studies have found a positive correlation between the degree of anteversion of the hip and the severity of DJD.[24] Any of these conditions may result in maldistribution of loads across the hip joint followed by localized stress within the tissues. Early detection and treatment of these conditions in childhood is essential for the prevention of a major cause of secondary DJD.

Microtrauma, as well as major trauma, has been considered a reason for development and progression of DJD.[18] This microtrauma comes mainly from rapid repetitive loading; apparently as little as 20 minutes per day of loading over a period of several months can be a factor.[25] Small repetitive loads can cause deleterious effects on cartilage and subchrondal bone, thus leading to initial cartilaginous changes.[6] Several authors have reported the occurrence of DJD following years of minimal repetitive trauma resulting from occupational and sports habits.[6,16,18,25,26]

There seems no doubt that local mechanical factors are major determinants of DJD,[21] and that the major factor responsible for cartilage destruction is uneven distribution of pressure. Some investigators have found that initial lesions occur in the areas of the joint that are not submitted to intermittent loading.[24,27] Their contention is that nutrition of the cartilage is stimulated by intermittent compression of the joint surfaces and that insufficient compression deprives the cartilage of proper nutrition resulting in cartilaginous lesions.[24,27] On the other hand, overloading when the contact area is small has also been found to result in cartilaginous destruction. In either case, too little or too much loading of cartilage, mechanical factors relating to physical activity appear to be the essential for the progression of DJD.

If a large stress is produced in the normal subchondral bone, fractures of the trabeculae may occur. The bone responds by remodeling in a manner to resist further fracture. The trabeculae becomes thicker and more cross struts are developed.[7] This change in trabecular form and mass makes the subchondral bone stiffer. In turn, the area is reduced in its ability to absorb energy. This change is followed by greater stress within the cartilage causing it to deteriorate.[26] Radin and associates report that doubling the stiffness of the subchondral bone increases the peak stress in cartilage by 100%.[6] The cartilage loses its ability to spread the load and decrease stresses so that the stresses become more concentrated within the cartilage and bone. If the rate of fractures produced is greater than the healing process, localized bony collapse ensues. Cysts may then develop in these areas.[17,23,25] In general, a progression of joint wear is established. The bone reacts to stresses by overgrowth and development of cysts with continued deterioration of the articular cartilage. Continued activity enhances this progression.

By recognizing that repetitive loading can cause the initial lesion of DJD and definitely provide for a progression, properly regulated exercise programs may be developed. In such patients, perhaps jogging or other repetitive weight bearing activities should be replaced by exercises which still benefit the cardiovascular system but do not overload the hip joint.

Fracture

Bony weakness as a result of loss of bone mass (osteoporosis) has been cited as playing a major role in the pathogenesis of fracture of the hip.[28,29,30] Osteoporosis describes a condition in which bone mass is significantly less than that for a given age group and is low enough that spontaneous fractures can occur. This condition proceeds more rapidly following menopause in women than it does in men of the same age.

In the proximal end of the femur, bone seems to be reduced within the thickness of the cortex by resorption of its inner surface.[31,32] A decrease in the number and thickness of the trabeculae also is present.[32,33] During the first four or five decades of life, internal replacement and adjustment of the bone mass is maintained according to the mechanical loads placed upon it. As the body ages, however, its ability to adapt to environmental stress declines and osteoblastic deposition of bone proceeds at a slower rate.[32] Thus, with the aging process, the bone adjustive remodeling process seems to lose its effectiveness.

In the case of the hip, the force of the superincumbent weight is transmitted by way of the bending moment through the proximal end of the femur, with tension on the superior aspect and compression on the inferior aspect of the structure (Figure 5). Bone is weaker in tension than in compression. Therefore, when the bone is weakened by loss of mass, the superior part of the femoral neck may not be able to withstand the forces of the body weight and inertia. An example of this situation is that of an individual's stepping down from a curb and sustaining a fracture of the hip. Strong abductor muscle contraction may prevent fracture, but a weak muscle may not protect the area adequately and fracture results.[13]

Although the causes of osteoporosis are numerous, and not well understood, the lack of mechanical loading seems to be a major factor.[34,35] The lack of physical activity, particularly weight bearing, appears to be related to loss of bone mass,[24] with immobilization resulting in local osteoporosis.[28] Since the strength of bone adapts to the loads placed upon it, inactive individuals have weaker bones. This fact, coupled with loss of estrogen in women later in life, results in further weakening of the bone.

Outstanding among suggested preventative measures for loss of bony mass is physical activity.[30,31,36–38] Continued physical activity from childhood seems to prevent or delay loss of bone density, allowing the normal skeleton to become denser and stronger.[41,42] As the person ages, continued activity can help the bone to maintain its density, or at least retard the loss of bone. Pronounced inactivity accelerates the age-related bone loss.[37] Therefore, older individuals should be strongly encouraged to maintain an exercise program, including standing and walking, that provides loading to the hip area. Aggressive physical therapy helps prevent immobilization and provides appropriate activity programs for those with osteoporosis, or for the elderly and handicapped with the potential for this disease.

CONCLUSION

The two previous sections of this article presented the extremes of how hip disorders may be caused, prevented, and treated by mechanical means. Degenerative joint disease and osteoporosis resulting in fractures of the proximal end of the femur are similar in that they affect patients of the same age groups.[38] Several studies, however, report that these two disorders are almost mutually exclusive, as patients with fracture of the femoral neck and osteoporosis seldom have DJD.[30,33,39,40] They are similar in that the amount of loading of the hip joint from physical activity at different times during the life of an individual has an important bearing on the etiology of both DJD and osteoporosis.[30] Those individuals who spend their lives in physical toil or athletics are less likely to have osteoporosis, but may have a greater tendency toward DJD. Those who are sedentary tend to have osteoporosis and do not have DJD. Understanding the mechanical aspects of DJD and osteoporosis are essential for the health professional to properly guide the patient between the two extremes of too much loading and too little loading of cartilage and bone.

REFERENCES

1. Johnston RC: Mechanical considerations of the hip. Arch Surg 107:411–417, 1973

2. Kempson GE: Mechanical properties of articular cartilage. In Freeman MAR (ed): Adult Articular Cartilage, ed 2. Kent, England, Pitman Medical Publishing Co Ltd, 1979, pp 333–414

3. Frost HM: Orthopedic Biomechanics. Springfield, Ill., Charles C Thomas, 1973

4. Weightman B, Kempson GE: Load carriage. In Freeman MAR (ed): Adult Articular Cartilage, ed 2. Kent, England, Pitman Medical Publishing Co Ltd, 1979, pp 291–331

5. Frankel VH, Nordin M: Basic Biomechanics of the Skeletal System. Philadelphia, Lea and Febiger, 1980, pp 149–177

6. Radin EL, Paul IL, Rose IM, et al: The mechanics of joints as it relates to their degenerations. In AAOS Symposium on Osteoarthritis. St. Louis, CV Mosby Co, 1976, pp 34–43

7. Pugh J: Biomechanical aspects of osteoarthritic joints: mechanisms and noninvasive detection. In Ghista DN (ed): Osteoarthromechanics. Washington, Hemisphere Publishing Corp., 1982, pp 161–191

8. Muckle DS, Bently G, Deane G, Kemp FH: Basic sciences of the hip. In Muckle DS

(ed): Femoral Neck Fractures and Hip Joint Injuries. New York, John Wiley & Sons, 1977, pp 1–54

9. Swanson SAV: Friction, wear and lubrication. In Freeman MAR (ed): Adult Articular Cartilage, ed 2. Kent, England, Pitman Medical Publishing Co Ltd, 1979, pp 415–457

10. Evans FG, Pedersen HE, Lissner HR: The role of tensile stress in the mechanism of femoral fractures. J Bone Jt Surg 33A: 485–501, 1951

11. Hall MC: The trabecular patterns of the neck of the femur with particular reference to changes in osteoporosis. Canad MAJ 85:1141–1144, Nov 18, 1961

12. Morris JM: Biomechanical aspects of the hip joint. Orthop Clin of Amer 2:33–54, 1971

13. Singleton MC, LeVeau BF: The hip joint—structure, stability, and stress. Phys Ther 55:957–973, 1975

14. McLeish RD, Charnley J: Abduction forces in the one-legged stance. J Biomechanics 3:191–209, 1970

15. Denham RA: Hip mechanics, J Bone Jt Surg 41A:550–557, 1957

16. Howell DS: Aging in osteoarthritis: osteoarthrosis or degenerative joint disease. In Kay MMB, Galpin J, Makinodan T (eds): Aging, Immunity and Arthritic Disease. New York, Raven Press, 1980, pp 135–150

17. Freeman MAR, Mechim G: Aging and degeneration. In Freeman MAR (ed): Adult Articular Cartilage, ed 2., Kent, England, Pitman Medical Publishing Co Ltd, 1979, pp 487–543

18. Moskowitz RW: Clinical and laboratory findings in osteoarthritis. In McCarty Jr. DJ (ed): Arthritis and Allied Conditions. Philadelphia, Lea and Febiger, 1979, pp 1161–1180

19. Salter RB: Textbook of Disorders and Injuries of the Musculoskeletal System. Baltimore, Williams and Wilkins, 1970

20. Hoagland FT: Osteoarthritis of the hip: etiologic factors and preventive measures. In AAOS Symposium on Osteoarthritis. St. Louis, CV Mosby Company, 1976, pp 66–79

21. Sokoloff L: Pathology and pathogenesis in osteoarthritis. In McCarty DJ JR (ed): Arthritis and Allied Conditions. Philadelphia, Lea and Febiger, 1979, pp 1132–1154

22. Salter RB: Disorders of children leading to degenerative arthritis of the hip in adulthood. In Cruess RL, Mitchell NS (eds): Surgical Management of Degenerative Arthritis of the Lower Limb. Philadelphia, Lea and Febiger, 1975, pp 33–41

23. Ferguson AB JR: The pathology of degenerative arthritis. In Cruess RL, Mitchell NS (eds): Surgical Management of Degenerative Arthritis of the Lower Limb. Philadelpha, Lea and Febiger, 1975, pp 3–10

24. Reikeras O, Hoiseth A: Femoral neck angles in osteoarthritis of the hip. Acta Orthop Scand 53:781–784, 1982

25. Radin EL: Mechanical aspects of osteoarthritis. Bull Rheum Dis 26:862–865, 1975–76

26. Radin EL, Paul IL, Rose RM: Role of mechanical factors in the pathogenesis of primary arthritis. Lancet: 519–521, March 4, 1972

27. Harrison HMH, Schajowicz F, Trueta J: Osteoarthritis of the hip, a study of the nature and evolution of the disease. J Bone Jt Surg 35B:598–626, 1952

28. Sissons HA: Osteoporosis and osteomalacia. In Ackerman LV, Spjut HJ, Abell MR (eds): Bones and Joints. Baltimore, The Williams and Wilkins Company, 1976, pp 25–38

29. Smith R: Metabolic aspects. In Muckle DS (ed): Femoral Neck Fractures and Hip Joint Injuries. New York, John Wiley & Sons, 1977, pp 120–148

30. Foss MVL, Byers PD: Bone density, osteoarthrosis of the hip, and fracture of the upper end of the femur. Ann Rheum Dis 31: 259–263, 1972

31. Raisz LG: Osteoporosis. J Amer Geriat Soc 30:127–138, 1982

32. Bogumill GP, Schwamm HA: Orthopedic Radiology. Philadelphia, WB Saunders, 1984, pp 241–296

33. Progrund H, Rutenberg M, Makin N, et al: Osteoarthritis of the hip joint and osteoporosis. Clin Orthop Rel Res 164:130–135, 1982

34. Hogue CC: Injury in late life: II Prevention. J Amer Geriat Soc 30: 276–280, 1982

35. Radin EL: Mechanical aspects of fractures and their treatment. In De Luca HF, Frost HM, Jee WSS, et al (eds): Osteoporosis: Recent Advances in Pathogenesis and Treatment. Baltimore, University Park Press, 1981, pp 191–199

36. Sandler RB: Etiology of primary osteoporosis: an hypothesis. J Amer Geriat Soc 24:209–213, 1978

37. Gordon GS, Vaughn C: Clinical management of the osteoporoses. Acton, Mass, Publishing Sciences Group, Inc, 1976

38. Gordon GS: If preventable why not prevented? Western J Med 133: 331–333, 1980

39. Weintraub S, Papo J, Ashkemazi M, et al: Osteoarthritis of the hip and fractures of the proximal end of the femur. Acta Orthop Scand 53: 261–264, 1982

40. Solomon L, Schnitzler CM, Browett JP: Osteoarthritis of the hip: the patient behind the disease. Ann Rheum Dis 41: 118–125, 1982

41. Overton TR, Hangartzer TN, Heath R, et al: The effect of physical activity on bone: gamma ray computed tomography. In De Luca HF, Frost HM, Jee WSS, et al (eds): Osteoporosis: Recent Advances in Pathogenesis and Treatment. Baltimore, University Park Press, 1981, pp 147–158

42. Smith EL, Reddan W: Physical activity—a modality for bone accretion in the aged. Amer J Roent 126–129, 1976

Evaluation of the Patient with Hip Pain

David S. Caldwell, M.D.
John R. Rice, M.D.

ABSTRACT. The most common complaint encountered in patients with hip disease is pain. A variety of disorders may cause pain, perceived by the patient to involve the hip. While frequently this is due to an articular disorder, disorders involving nonarticular structures including periarticular tissue (tendon, bursa) and viscera of the abdomen and pelvis likewise must be considered. The history, physical examination, and selected studies are of utmost importance in determining the specific cause of hip pain. Once the cause is determined, a specific therapeutic program can then be designed, influenced not only by the specific disorder, but also by the age, general health, functional expectations and motivation of the patient.

By virtue of its anatomy, the hip is an extremely stable joint, capable of a remarkable range of motion. Likewise, it plays a major role in locomotion and in bipeds is subjected to considerable stress, frequently receiving the brunt of trauma. In direct injury, or in local or systemic diseases with hip involvement, the resultant dysfunction produces not only severe pain, but also greatly limits the activities of the patient.

HISTORY

The most common complaint in the patient with hip disease is pain. Rarely, a characteristically painless disorder such as congenital dislocation, congenital coxa vara, or neurogenic arthropathy may occur. Owing to multiple innervations of the hip (obturator,

David S. Caldwell, Assistant Professor of Medicine, Division of Rheumatology and Immunology, P.O. Box 2978, Duke University Medical Center, Durham, NC 27710. Dr. Caldwell serves on the North Carolina Board of Physical Therapy Examiners.

John R. Rice, Assistant Professor of Medicine, Division of Rheumatology and Immunology, P.O. Box 3383, Duke University Medical Center, Durham, NC 27710.

sciatic, and femoral nerves), pain may be experienced in the groin region in addition to the lateral, anterior, or posterior aspects of the hip. In order to reduce the potential number of diagnostic possibilities accounting for the patient's pain, the examiner should carefully question the patient on the precise localization of his pain. More often than not, the patient complaining of "hip pain" will localize this to the buttock. While true hip disease may cause buttock pain, a variety of other conditions must be considered in this situation, particularly lumbar nerve root irritation or sacroiliitis. The patient also may complain of thigh pain and, on further questioning and examination, this may be more specifically localized to the true hip or inguinal area. If not, other entities including herniated lumbar disc, particularly at the L3-4 level, trochanteric bursitis, abductor tendonitis, psoas abscess, and meralgia paresthetica (lateral femoral cutaneous nerve entrapment) must be considered. While inguinal or groin pain strongly suggests a hip source to the examiner, this complaint as an isolated symptom is rare in the authors' experience.

Pain secondary to internal hip disease may also refer to the knee, following Hilton's law which states that the nerve trunk supplying a group of muscles that move a joint also supplies nerves to the skin over the insertion of the same muscle group and to the interior of the joint.

While specific localization of the pain may be sufficient to incriminate the hip joint as the cause, two additional complaints are particularly useful diagnostically. First, pain on weight bearing or a pain aggravated by weight bearing is typical of most diseases involving the articular surfaces. This pain usually results in a limp which may be the first symptom mentioned by the patient. Occasionally, a limp may be present in the absence of pain. Second, patients with intrinsic hip disease usually have limited movement. The most significant complaint is the loss of the ability to rotate the leg into abduction when putting on a sock or stocking. Weakness and disability relating to hip dysfunction are often elicited as well.

While hip pain usually is aggravated by weight bearing, and relieved by rest, pain during the night may be indicative of the severity of hip disease, i.e., a direct relationship between intensity of nocturnal pain and disease intensity. Neoplasms involving the hip and tuberculosis of the hip also may be worse at night. Hip pain caused by rheumatoid arthritis may be significant on awakening and diminish somewhat with activity. Pain in the hip region made worse by lying on the ipsilateral side indicates trochanteric bursitis, while pain aggravated by sitting may be a symptom of ischial bursitis. Acute pain in the hip region, associated with positioning of the leg in slight flexion, abduction, and external rotation for relief, usually is due to a distended joint capsule, i.e., infectious arthritis, acute inflammatory synovitis.

Lastly, many nonarticular problems may cause a painful hip. Most often such a referred pain pattern would be a result of lumbosacral or visceral disease. Questioning the patient regarding prior illnesses or injury is of utmost importance, particularly if there is a suggestion from the patient's description of the pain that a cause other than true hip disease may be present. The presence of systemic symptoms such as fever, weight loss, lymphadenopathy, or the history of a prior malignancy with potential for bone metastasis are particularly ominous and require an expedient and thorough evaluation. The middle-aged mildly obese patient with a prior history of sciatica who presents with buttock pain radiating into the posterior thigh should be carefully evaluated for nerve root irritation. The elderly female patient with buttock pain undoubtedly has osteoporosis and may have sustained a vertebral compression fracture or possibly an occult sacral fracture to account for the symptoms. Pelvic inflammatory disease, renal disease, and vascular insufficiency may all bring the patient to the physician with complaints referrable to the region of the hip. The medical history is of utmost importance, frequently providing insight into the etiology of hip pain and thereby directing the physical examination and subsequent diagnostic studies (Table I).

TABLE I : EVALUATION OF THE PATIENT WITH HIP PAIN : HISTORY

- where is the pain located ?
- is pain affected by weight bearing?
- is there a limp ?
- is there limitations of movement in activities of daily living ?
- what are the aggravating and relieving factors ?
- is there another pre-existing disease in this patient that could cause hip pain ?

PHYSICAL EXAMINATION

Thorough examination of the hips is performed by inspection, palpation, observation of range of motion and movement, and muscle testing, in erect, prone, and supine positions. While a complete physical examination is important, particular emphasis should be given to the lumbar spine, lower extremities, abdomen, and pelvis to exclude nonarticular causes of hip pain. Special tests for the hip also may be important. Notation of the patient's body habitus, posture, nutritional state, and previous operative scars frequently will provide clues to the etiology of hip pain.

In the gait of a normal, healthy individual, abductor muscles of the weight-bearing extremity contract to maintain a level pelvis or slight elevation of the non-weight-bearing side of the pelvis. When the hip is painful, the body may tilt toward the involved hip on weight bearing to maintain balance and avoid abductor muscle contraction with resultant painful muscle spasm. This is an antalgic gait. Because of nonuse, the abductors of the involved hip may then weaken and become unable to hold the pelvis level when weight is borne by the involved hip. The pelvis then drops on the side opposite the involved hip, producing a Trendelenburg gait (gluteal, gluteus medius, or abductor gait). The patient may in effect lean away from the abnormal hip. If both abductors are weak, a waddling (duck) gait results. This may be seen in diseases affecting both hips. Complete ankylosing of the hip forces the patients to "swing" the leg from the lumbar spine and hence a "swinging gait" may be seen.

On further inspection, one may note that the patient assumes a position of hip flexion, which is one of the most common findings suggesting hip abnormality. The flexed hip relaxes the articular capsule which may be distended with effusion, and relieves painful muscle spasm. In the supine position this may be erroneously reduced by anterior arching of the spinal column, resulting in anterior pelvic tilt and exaggerated lumbar lordosis, allowing the thigh to contact the examining table. Flexion of the opposite hip by the examiner flattens the lumbar spine and unmasks the flexion deformity (Thomas test). Passive extension of the flexed hip presses the femoral head firmly into the acetabulum, producing pain. Apparent discrepancies in leg length actually may be caused by lateral tilting of the pelvis associated with abduction or adduction deformities of the hips. Actual clinical or radiographic measurement will reveal equal leg lengths or less than 1 cm difference. Asymmetry in cutaneous folds in the gluteal region should be noted, as well as positioning of the lower extremity. A fracture of the femoral neck classically produces external rotation.

Because the hip joint is deeply located, overt swelling usually is non-detectable by palpation and localized heat or erythema in the region of the hip usually is an indication of overlying soft tissue inflammation. Palpation may reveal other conditions simulating hip disease. Localized anterior swelling lateral to the femoral artery may be a distended iliopectineal bursa, possibly communicating with the hip (15% of cases). This may be confused with a femoral hernia; however, the latter is usually medial to the artery. Aneurysms of the femoral artery, saphenous vein varix, psoas abscess, inguinal adenopathy, and rectus femoris muscle rupture all may appear as swelling anterior to the hip joint. Discrete localized tenderness on palpation, often elicited with the patient's direction, may provide a diagnosis of bursitis. Trochanteric bursitis causes localized tenderness and occasionally swelling over the greater trochanter of the femur. The pain usually is aggravated by active abduction and external rotation of the hip carried out against resistance. Tenderness over the ischial tuberosity may indicate ischiogluteal bursitis (Weaver's bottom), which rarely is associated with detectable swelling, described as a circumscribed bulge with a doughy consistency. Percussion of the heel of the foot with the examiner's fist (anvil test), as well as trochanter-to-trochanter pressure, may elicit pain in the hip in patients with true hip disease. These maneuvers may be particularly useful when the patient is unable to bear weight.

Normal hip anatomy provides for a wide range of motion permitting flexion, extension, adduction, abduction, rotation, and circumduction. Rotary movement of the femoral head within the acetabulum is provided by the angulation between the neck and shaft of the femur which partially converts angular movements into rotary movements. Normal ranges of motion vary and loss of motion may be determined best by comparing the affected hip to the uninvolved hip. Hip flexion may be assessed with the knee extended or flexed to 90 degrees. If the knee is extended, hip flexion will not exceed 90 degrees owing to tightening of the hamstrings. Extension of the hip is best performed with the patient lying on the opposite side. Except for this part of the examination, range of motion is determined most effectively with the patient supine.

Abduction may be evaluated with the knees either flexed or extended, by actively or passively moving the leg away from the neutral position (midline of the body). Flexion of the hip will increase the degree of abduction. Adduction is performed by crossing one leg over the other. Rotation is assessed by rolling the extended leg from one side to the other or, with the knee flexed, swinging the foot inward (external rotation) or outward (internal rotation). Internal rotation, though frequently impaired in older patients with no apparent hip disease, is the first range of motion to be impaired in hip disease.

The most common deformity of the hip joint is a flexion contracture with concomitant knee flexion; hip extension deformity is not usually a clinical problem. Adduction contractures may occur, particularly in diseases characterized by prolonged inflammation such as rheumatoid arthritis. This may result in difficulty with ambulation as well as sexual function and personal hygiene. Lastly, an assessment of muscle strength is important to determine possible neurogenic factors contributing to pain, as well as to assess muscle atrophy resulting from nonuse. Prime movers of the hips in flexion, extension, abduction, adduction, lateral rotation, and medial rotation are each assessed by the examiner. Measurement of thigh circumference bilaterally to assess muscle atrophy may be useful (Table II).

GLOSSARY OF SPECIAL TESTS TO EVALUATE THE HIP

1. Heel to knee test or fabere sign (Patrick's test)—Test involves several motions of the hip simultaneously to detect hip joint involvement. Motions utilized include *F*lexion, *AB*duction, *Exter-*

TABLE II : EVALUATION OF THE PATIENT WITH HIP PAIN— PHYSICAL EXAM

1. Inspection: gait, body habitus, posture, operative scars, nutritional status

2. Palpation: swelling, tenderness

3. Observation: joint mobility, range of motion, deformity

4. Muscle strength testing

5. Special physical maneuvers

nal *R*otation and at the completion of the test, *E*xtension (fabere— not an eponym and hence not capitalized). The hip and knee on the tested side are flexed so that the heel lies beside or on top of the opposite extended knee. The hip being tested is then abducted and externally rotated as far as possible. Pain, spasm or limited motion in the tested hip constitutes a positive test. This provides a general indication of hip disease. Lifting of the opposite hip from the examining table constitutes a positive test as well.

2. Trendelenburg test—In the normal weight-bearing extremity, abductors contract to elevate the opposite side of the pelvis to maintain balance. In the diseased hip, abductors may be painful or weakened and thus unable to raise the opposite hip. A positive Trendelenburg test results when the opposite hip drops as weight is borne by the painful hip.

3. Ober test—With the patient on his side, affected side up, the uppermost extremity is extended and abducted at the hip with the knee flexed. The thigh is supported by the examiner. After releasing the involved thigh, the knee should drop. If it does not, the test is positive and indicates a contracture of the iliotibial band. If there is hip flexor muscle contracture, this test also produces hip pain.

4. Jansen's test—This tests the ability of the patient to cross his legs, resting one ankle on the opposite knee, and is considered positive if the patient is unable to complete the maneuver. The test frequently is positive in patients with osteoarthritis.

5. Erichsen's sign—Compression of the iliac bones towards one another with the patient on his side will elicit pain if sacroiliac disease is present. This may help to rule out a primary hip disorder.

DIAGNOSTIC STUDIES

Findings from the history and physical exam may then be utilized to select appropriate diagnostic studies to more specifically determine the cause of the patient's painful hip. An incomplete history and/or physical examination may belie the potentially serious etiology of hip pain and necessitate more studies which are not only costly and time consuming, but also may delay much needed therapy. Initial studies should include radiographs of the hip, including an anteroposterior view of the pelvis to compare the affected side with the unaffected side. The presence of osteophytes and sclerosis may suggest osteoarthritis, while periarticular osteoporosis and joint-space narrowing with axial migration of the femoral head would be consistent with rheumatoid disease. Infection is usually associated with rapid destruction of both the acetabulum and femoral head. Incongruous joint surfaces with secondary degenerative changes suggest underlying congenital hip dysplasia, Legg-Calvé-Perthes disease, or slipped capital femoral epiphysis. Helpful laboratory studies include a complete blood count, westergren sedimentation rate, and rheumatoid factor. When the diagnosis is still uncertain, additional studies may include bone scanning, tomography, arthrography, and occasionally hip aspiration for crystal analysis or culture or both.

CAUSES OF HIP PAIN

The most common disorder affecting the synovium is inflammation. Inflammation of hip synovium, or hip synovitis, may be a local process or part of a systemic illness. In the case of the latter, the hip rarely is involved first, thus making the diagnosis of hip pain more apparent when there is a pre-existing disease. For example, in rheumatoid arthritis (RA), the hip is rarely involved with the initial onset of this disease; eventually, however, radiographic evidence of hip disease develops in approximately half the patients. Five percent of all patients with RA will eventually develop destruction of the femoral head and remodeling of the acetabulum, forming the so-called "otto" pelvis or protrusio acetabuli. Loss of internal rotation on physical examination correlates best with radiographic findings. Hip involvement in juvenile rheumatoid arthritis occurs earlier and may be a more rapidly destructive process. All too often the hip may be a major source of disability in children with juvenile rheumatoid arthritis.

Synovitis of the hip occurring as an isolated finding may have several notable causes including crystal induced disease, infection, and a temporary, nonspecific inflammation of the hip seen in children called transient synovitis of the hip. Acute gout is primarily a disease of the lower extremity, but, even so, the hip rarely is involved initially or even during the course. A prior diagnosis of gout or a prior history of symptoms in other joints compatible with gout may provide a clue to the etiology of the patient's hip pain. Confirmation of this diagnosis, as well as the diagnosis of pseudogout, requires crystal analysis of joint fluid. Infectious synovitis also requires confirmation by joint fluid examination and culture. While far more common in children than adults, septic arthritis should always be considered in the patient with acute hip pain and fever. A more indolent course associated with persistent pain and extensive destructive changes by x-ray suggests tuberculosis.

While inflammatory diseases of hip synovium, if persistent, may ultimately damage the articular cartilage with resultant joint incongruity and pain, osteoarthritis begins as a disruptive process on the weight bearing contact surface of the articular cartilage. While inflammation may contribute to the course of this illness, it does not appear to be the precipitating event. Osteoarthritis of the hip is one of the most common joint abnormalities in patients over 40 years of age. Developmental anomalies of the hips such as congenital hip dysplasia and slipped capital femoral epiphysis may be predisposing factors in as many as 80% of patients with osteoarthritis of the hip, particularly those with early onset of this disease. Symptoms of osteoarthritis in the hip include pain in the inguinal or groin area and occasionally trochanteric and lateral buttock or proximal thigh pain. In 20% of patients the presenting complaint may be distal thigh or knee pain. Early in the patient's course, flexion may be pain-free, but extension and rotation are painful. Internal rotation is the first motion lost followed by compromise of extension, abduction, and hip flexion. While the patient with hip pain may have osteoarthritis elsewhere suggesting that disorder as the etiology of the painful hip, other causes must be considered. In the absence of any other osteoarthritis, diagnosis may depend on x-ray and radiologic findings that usually correlate well with clinical symptoms.

Disorders affecting the juxta-articular bone are variable and include stress fractures, avascular necrosis, and malignancy. Stress fractures occurring in the conditioned individual more frequently involve the tibia or foot, but persistent hip pain even in the absence of x-ray change should suggest the need for additional study, particularly bone scan and tomography. Patients with osteoporosis frequently sustain compression fractures of the vertebral bodies causing back pain. Less frequently appreciated in this clinical setting is spontaneous fracture occurring in the pelvis, presenting with low back, buttock or true hip pain. There is often no antecedent trauma or the immediately preceding event was quite minor, i.e., lifting a suitcase. Here as well, bone scan and subsequent tomography may prove diagnostic. The fracture may not be readily apparent on plain films; however, the bone scan will show increased activity in the fracture line and the nature of this activity can then be clearly defined by tomography. Avulsions and ruptures of muscles about the hips may result from trauma and produce considerable hip pain. History is usually apparent and examination often reveals a firm mass in the anterior thigh with hematoma formation.

Avascular necrosis involves the femoral head more frequently than any other site. It may occur for a variety of reasons including a complication of a fracture, most often a subcapital fracture of the femoral neck; a complication of long term steroid therapy; in association with chronic alcoholism; or as an idiopathic process (Chandler's disease). Pain usually begins insidiously and a slight limp actually may be the first manifestation of disease. This condition is quite easily distinguished from other entities radiographically at least in its early states. Tomography may define the characteristic changes more clearly.

Malignancy may invade the acetabulum or proximal femur producing typical hip pain. Most often the tumor is metastatic. The common primary sites are lung, kidney, thyroid, breast, prostate, and gastrointestinal tract. Pain resulting from metastatic involvement of the hip may precede plain radiographic evidence of destruction. If the clinical setting suggests such a possibility (patient with prior history of malignancy presenting with hip pain), bone scan, tomography, and possible biopsy of the abnormal area may be required to establish a diagnosis. Primary bone tumors including osteochondromas, chondrosarcomas, and fibrosarcomas, as well as tumors of muscle (cavernous hemangiomas, rhabdomyosarcomas, and rhabdomyomas), may cause hip pain if involving the immediate area. Paget's disease, while not a malignancy, frequently involves the pelvis and femur and may become symptomatic particularly when the joint is involved. The radiographic appearance may be confused with malignancy.

Inflammation or injury to periarticular tissues may be sufficient to cause limitation of movement and pain about the hip. Numerous bursae have been described about the hip but only three are of major importance; trochanteric, iliopectineal, and ischiogluteal. These have been reviewed earlier. Hamstring, abductor, adductor, and rotator tendons may become inflamed at their insertions into bone— so called enthesopathy. This may result from mechanical irritation or be a manifestation of the spondyloarthritic disorders including ankylosing spondylitis, psoriatic arthritis and Reiter's disease. Both inguinal and femoral herniae may produce groin pain and limitation of motion. Regional disorders likewise may bring the patient with a complaint of hip pain to the physician. "Snapping hip," an indication of a taut iliotibial band slipping over the greater tuberosity, may occur as the patient flexes and internally rotates the thigh. If the underlying bursa is inflamed this snapping sensation may be painful. Pain may also result from inflammation or tightness of the fascia lata as a result of overuse or nonuse. Pain is described over the low back, lateral hip, and thigh, with radiation down the lateral thigh to the lateral knee region. Pain may be noted on arising and with prolonged walking. Examination is performed with the patient on his side, on the edge of the table facing the examiner with the symptomatic side up. The straight leg is then drawn over the edge of the table and downward pressure placed on the ankle. This draws the fascia lata taut, producing pain, and occasionally dimpling over the fascia due to adhesions.

Symptoms of numbness and tingling about the hip suggest the possibility of an entrapment neuropathy. The abdominal cutaneous nerve may produce symptoms in the abdominal wall just above the inguinal area while the obturator nerve if entrapped may cause pain and paresthesias in the groin traveling down the inner aspect of the thigh, aggravated by hip motion. Meralgia paresthetica, entrapment of the lateral femoral cutaneous nerve, causes intermittent paresthesia, hypesthesia, or hyperesthesia over the upper anterolateral thigh. Though it mimics other conditions about the hip and thigh, hip range of motion is normal.

Polymyalgia rheumatica, an inflammatory disorder in older patients, presents with hip and shoulder girdle myalgias. Rarely would only the hip girdle be involved. Pain and stiffness are severe, but passive range of motion is usually not affected. An elevated sedimentation rate is essential for diagnosis. Fibromyalgia, characterized by musculoskeletal pain and diffuse trigger points, typically involves the pelvic girdle with trigger points about the greater trochanters and low back. Joint range of motion is not affected and studies are normal or non-diagnostic.

Lastly, a variety of disorders may refer pain to the hip region, thus simulating true hip disease. Osteoarthritis of the lumbar facet joints may produce buttock pain. L1–2 disc lesions may produce hip pain while disc lesions at L3–4 often produce radicular pain over the upper anterior thigh. Arteriosclerosis of the iliac arteries with partial or complete occlusion may produce hip and thigh pain on exertion which is relieved with rest (intermittent claudication). Femoral vein phlebitis may cause groin pain. An abscess of the psoas muscle may refer pain to the hip, and eventually cause limitation of motion. Several visceral abnormalities, owing to their location, may cause hip pain. These include renal colic, and pelvic inflammatory disease (Table III).

MANAGEMENT OF THE PATIENT WITH HIP PAIN

Important considerations in planning the management of the patient with hip disease include the patient's age, general health, life expectancy, motivation, type of hip disease, and functional expectations. Obviously, the younger patient with devastating hip disease in the absence of any other significant illness affecting function would likely be managed more aggessively. General therapeutic modalities apply to the patient with hip pain and include rest, medications, physical therapy, and when appropriate, weight reduction. Rest may be local, directed to the involved area by using canes or crutches, or be generalized for severe pain, i.e., bedrest. Medications, depending on the etiology of pain, may include nonsteroidal anti-inflammatory drugs, analgesics, muscle relaxants, or injectables such as soft tissue injections for bursitis or intra-articular injections for inflammatory synovitis. Physical therapy may include application of cold or heat for acute or chronic conditions, respectively, as well as a variety of exercises to maintain or restore function and strength. A stationary bicycle used without tension will help to maintain motion and muscle strength. Active and assisted range of motion exercises may help prevent deformities. Avoidance of certain positions such as sleeping on the side of

TABLE III: CAUSES OF HIP PAIN

1. Disorders affecting the synovium
2. Disorders affecting the articular surface
3. Disorders affecting juxta-articular bone
4. Disorders of periarticular soft tissue
5. Disorders of regional soft tissue
6. Disorders referring pain to the hip

the affected hip should be stressed and the patient taught to sleep supine, prone, or on the unaffected side. Use of cane or crutch in the hand opposite the affected hip decreases the force on the painful hip and hence reduces pain. Patients who may require two canes or crutches are invariably surgical candidates, while the patient who will not use or does not need an assistive device likely will not require surgical management. Definitive therapy for the patient with hip pain is obviously dependent on determining the precise etiology. That in turn depends on a careful thorough history, physical examination, and directed x-ray as well as laboratory evaluation.

SUGGESTED READING
FOR ADDITIONAL INFORMATION

Polley, HF, Hunder GF: Rheumatologic Interviewing and Physical Examination of the Joints. Second edition. Philadelphia, WB Saunders Company, 1978, 286 pp.

Sheon, RP, Moskowitz, RW, Goldberg VM: Soft Tissue Rheumatic Pain: Recognition, Management, Prevention. Philadelphia, Lea and Febiger, 1982, 302 pp.

Katz, WA: Rheumatic Diseases, Diagnosis and Management. Philadelphia, JB Lippincott Company, 1977, 1057 pp.

Beary, III, JF, Christian, CL, Sculco, TP: Manual of Rheumatology and Outpatient Orthopedic Disorders. Diagnosis and Therapy. Boston, Little, Brown and Company, 1981, 366 pp.

Kelley, WN, Harris, Jr. ED, Ruddy, S, Sledge, CB: Textbook of Rheumatology. Second edition. Philadelphia, WB Saunders Company, 1985, 1972 pp.

Swezey, RL: Arthritis: Rational Therapy and Rehabilitation. Philadelphia, WB Saunders Company, 1978, 242 pp.

Hip Fracture:
An Epidemic Challenging
the Physical Therapist

Frank Hielema, Ph.D., P.T.

ABSTRACT. An epidemic of hip fracture is challenging the health care resources of the developed nations. The growth in the epidemic results from both an increase in volume of the population at risk and an increase in the fracture rate within the population at risk.

This paper examines the antecedents of hip fracture: osteoporosis and falls. While some loss of bone mass occurs normally with advancing age, risk for osteoporosis is especially high among white women. Other risk factors include deficiency of calcium and Vitamin D in the diet, alcohol consumption, cigarette smoking, and low levels of physical activity. Among the young elderly, environmental factors contributing to tripping and slipping appear to be major contributors to falling. In the older elderly, however, poor health is more often implicated as a cause for falling. Health conditions, associated with falling include the use of sedative drugs, dizziness, cerebral ischemia, and functional disabilities resulting from stiff arthritic joints or weakness. Recent research in postural sway offers exciting clues for understanding the reasons for the increasing rate of falls with advancing age.

The epidemiology of hip fracture is also reviewed here. The risk factors for hip fracture understandably roughly parallel those for osteoporosis. As patients with hip fracture are usually referred to physical therapy, our clinicians must be prepared to meet the challenge posed by this epidemic. Physical therapists are also frequently in an excellent position to provide advice, guidance or intervention relative to osteoporosis and falls.

Frank Hielema, Director of Rehabilitation and Research, Hillhaven Convalescent Center and Adjunct Instructor, Department of Epidemiology, School of Public Health, University of North Carolina, Chapel Hill, NC 27514, Chairman, North Carolina Governor's Council on Physical Fitness and Health.

The author wishes to acknowledge the assistance of Brigitte Scek in the preparation of this manuscript.

49

The industrialized nations of the world, which have achieved life expectancies beyond 70 years of age, are facing an epidemic of hip fracture which has significant implications for the health care systems of these countries. Patients with this condition are usually referred for physical therapy and thus, this epidemic presents a tremendous challenge to our profession's knowledge, skills, and manpower resources.

This paper will briefly discuss osteoporosis which is the underlying cause of hip fracture. Falling, the event which marks the onset of the fracture in the elderly, will be examined. Attention will be paid to the epidemiology of hip fracture.

The incidence of hip fracture has been found to be increasing in recent years, even after age and sex standardization accounts for the increasing numbers of elderly women in the population structure.[1-3] Wallace has offered two hypotheses to explain the increase.[3] First, as the elderly in industrialized societies have had to live on retirement incomes which have not kept pace with inflation in the past 15 years, they have had to limit their purchases of nonessential food items, including milk and other dairy products rich in calcium. Calcium deficiency, as it will be seen later, is clearly associated with osteoporosis. His second hypothesis relates to decreasing activity levels among people of the industrialized nations over the past two decades, the result of greater mechanization of labor and improved social services.

In addition to this increase in rates per specific volume of population at risk, the population at risk is also growing rapidly in the industrialized nations. Public health and medical advances have allowed increasing numbers of people to survive the infectious disease epidemics. Progress has been made in prolonging the lifespan of individuals with chronic disease of a vital organ system into the seventh or eighth decade of life. The media confront us almost weekly with astounding projections for growth in the numbers of people over age 75 by the year 2000 or even 2025.

Given that limited resources are available to the health care system in terms of dollars, beds, and personnel, what are the implications? Already, England has experienced a decrease in orthopedic beds available for patients needing or desiring elective joint replacements because of the rise in demand for emergency treatment of hip fractures.[2] Treatment is expensive. Prior to the institution of Diagnosis Related Groups (DRG) in the United States, the average length of acute hospital stay among those over 65 years of age with hip fracture was 22.8 days.[4] Many of those discharged from the hospital require long term care placement due to their inability to resume independent living. In fact, one survey found that 15 percent of skilled nursing facility admissions in one area of the United States were for hip fracture.[5] Physical therapy and occupational therapy may be required for weeks or months in the nursing home or the patient's home. The patient who is fortunate enough to go directly home from the hospital may still require home health aide services for support.

For physical therapists, to meet the challenge of this epidemic, they must possess the knowledge that allows them to provide the appropriate services and the numbers of personnel who are actually *interested* in working with this traditionally unpopular clientele. We would all like to believe that physical therapy is important in the recovery from hip fracture; however, we do not yet have solid research evidence to guide us in achieving the balance between sufficient service of high quality (if indeed physical therapy is beneficial at all) and the costly overdosage of expensive treatment. Physical therapists, along with other health care researchers, must look for this evidence. Given that physical therapy is necessary for recovery from hip fracture, our educational institutions are responsible for insuring that *interested* personnel are available to meet the epidemic.

THE CULPRIT BEHIND THE EPIDEMIC: OSTEOPOROSIS

Osteoporosis is the villain responsible for the increased rate of certain fractures with advancing age, particularly in women. This condition is characterized by a decrease in bone mass or radiographic density which is a result of faster bone resorption than formation.[6,7] As many as 15 to 20 million Americans are estimated to suffer from osteoporosis. The condition, however, becomes of clinical consequence usually only as a result of a fracture involving one of three typical sites: vertebral bodies (T8 to L3 are the most often affected vertebrae), the distal radius (Colles' fracture), or the hip.[6]

Loss of bone mass occurs normally with advancing age. Women have lower density than men at early ages and suffer from an increased rate of bone loss following natural or surgical menopause.

White women, and particularly those of fair complexion, are at higher risk than black women. White men are also at higher risk than black men. Various nutritional factors have been associated with osteoporosis. Calcium deficiency in the diet has been strongly implicated as a causal factor. Alcohol consumption and dietary deficiency of Vitamin D and fluorides may also increase the risk of this disease.[6-8]

Cigarette smoking, already known to increase risk for lung cancer, heart disease, and bladder cancer, also appears to be associated with osteoporosis.[6] Of special interest to physical therapists is the association between inactivity and accelerated bone loss. Bone loss is increased with extended bed rest and immobilization. Weight bearing exercise has been found to decrease bone loss and even increase bone mass.[6] It has recently been discovered that among patients with anorexia nervosa, those who report a high level of physical activity are not as likely as those less active to suffer the loss of bone density and accompanying vertebral compression fractures.[9]

Osteoporosis can be treated, retarded, or perhaps even prevented. The most common therapies known to be beneficial include estrogen supplementation, dietary calcium and Vitamin D. Modest weight-bearing exercise, such as walking, has also been recommended as a preventive measure. Unfortunately, therapies for osteoporosis carry added risks for developing other health conditions. Estrogen therapy has been associated with an increased risk for endometrial cancer.[10] Excessive dietary calcium can increase the risk of urinary tract stones in persons at risk for this condition. Vitamin D can become toxic at high doses, and of course, initiating an exercise program involves an increase in risk for musculoskeletal injuries.[6,7]

Physical therapists should be alert to whether their patients with hip or other fractures associated with osteoporosis are receiving treatment for this bone disorder. If treatment is not being received, it would be appropriate to ask the patient if this issue has been discussed with the physician. If the answer is negative, the therapist should feel free to advise the patient to discuss treatment choices and their risks and benefits with the physician. Such an approach is in philosophical agreement with our nation's current embracement of the wellness movement which advocates increased patient knowledge regarding decreasing disease risks.

THE FALL: A QUESTION OF, "WHICH CAME FIRST, THE CHICKEN OR THE EGG?"

Invariably, in discussion of hip fracture, the following question arises: "I've heard it said that the fall results from the fracture. Is this so?" In other words, can a spontaneous hip fracture interrupt weight bearing so that the individual will collapse to the floor? It is certainly conceivable that, in the severely osteoporotic proximal femur, minimal stress such as weight bearing or torsion can produce a fracture; however, a review of the literature on the causes of falls does not implicate spontaneous fracture.[11-14] Therefore, we can conclude that the act of falling itself is a significant contributor beyond osteoporosis to the hip fracture problem. Consequently, research on falls may provide another route whereby intervention may be helpful in limiting this epidemic.

Falls are a significant problem among the geriatric population. Among the community based elderly, up to 40 percent may experience a fall within a given one year period.[11,14] A similar percentage of the institutionalized elderly appear to be at risk, but within a less active lifestyle.[11] Increasing age, living alone, and female gender are classical risk factors for falling,[11,14] but these variables may lose some of their strength of prediction after health status is controlled.[11]

For younger persons, environmental factors appear to be the main cause of falling.[11,14] These factors include obstacles in one's path, curbs, stairways, uneven pavement and slippery surfaces which an active individual faces as a daily challenge. Among the older elderly, however, health problems appear more often to be the cause. Many health problems are frequently mentioned,[11,13,14] although it is not clear for what proportion of the problem they might be responsible. Dizziness is an often mentioned cause.[11-14] Cerebral ischemia, resulting from arteriosclerotic cerebrovascular disease, postural hypotension, cardiac arrhythmia or other conditions, is frequently mentioned as a significant causal factor.[11,13] Other organic brain diseases, such as Alzheimer's disease, may affect areas of the brain which play a role in postural control.[14] Functional disabilities from stiff arthritic joints or weakness may be important.[11]

Recent research related to postural sway provides important clues to understanding the reasons for the increasing rate of falls with age. Sway has been found to increase with age and to be greater in women than in men of all ages.[15] Increased sway is present in persons who report falls because of balance loss, dizziness, or lack of postural control, (i.e., factors prevalent among the older elderly), but not in those who fall due to environmental obstacles. Further study has shown that postural sway is significantly correlated with decreased vibration sense in the lower extremities.[16] Impairment of proprioception, and decreased vision and vestibular sense, appear to be less significant in contributing to sway.[16]

Although falls are a significant and potentially medically serious problem for the elderly, not every episode leads to fracture or even significant injury. Because of variations in data collection among studies it is difficult to arrive at a reliable estimate of risk for injury or fracture among those who do fall. Among the elderly living at home, between eight and 40 percent of falls result in fracture of some bone.[11] The severity of injuries and fractures suffered is a function of the victim's sex and age, which we might conclude is the result of association of osteoporosis with these variables.

An all too frequent error is made in taking away the freedom of mobility of the individual who has experienced one or more falls. It is a gross error of judgement for health care providers to deny this right without first considering environmental and health factors which may have contributed to the fall, and then taking action to remove those factors. The *individual's* particular risk for minor and serious injury must be considered. The patient should be informed of the risk and, if he desires, be given an active role in the decision making process for decreasing risk. If for reasons of mental incompetence the patient is unable to participate in this decision making process, the right goes to the family or responsible party.

Exercises for improving balance and patient confidence that he will not fall have been described,[13] but the value of these has not been proven in controlled clinical trials.

THE EPIDEMIOLOGY OF HIP FRACTURE

Quickly it becomes apparent that the risk factors associated with hip fracture approximate those for osteoporosis. This is to be expected since osteoporosis is the underlying condition which frequently manifests itself clinically as hip fracture.

First of all, women are at much greater risk for this injury than men, with rates for women being 2.4 to 4.0 times higher than for men in various populations.[1,17] Hip fracture is strongly associated with increasing age in both men and women. In various studies the mean age of patients suffering this injury has ranged from 70 to 78 years.[1,17–21] Incidence increases rapidly in the last decades of life as demonstrated by the figures for one region of Britain.[22] In this sample, the incidence among women aged 70–74 years was 3.5 per 1000. This continued to increase to 6.3 per 1000 at 75–79 years, to 13.0 per 1000 at 80–84 years, 22.9 per 1000 at 85–89 years and to 32.8 per 1000 for women aged 90 to 94.

Considerable variation in hip fracture rates exists among countries,[1,23,24] and even within a single nation.[18,19] Geographic differences in incidence may be the result of biological differences among the populations,[1] or be a reflection of variation in life expectancy, race, availability of treatment for osteoporosis or other factors among groups. Hip fracture appears to be primarily an affliction of Western peoples.[20] Whites are definitely at greater risk than blacks, even when the longer life expectancy of whites is taken into account.[1,20,25,26]

The majority of investigators have found the predominant hip fracture type to be trochanteric,[17,21,27] but some have found the major type to be transcervical.[1] Analysis of incidence rates suggests that trochanteric fractures become more common in all persons over 80 years of age and are the preponderant type in men over 85 years.[1,28] Whether a trochanteric fracture presents as comminuted and unstable or not is related to the degree of osteoporosis present.[28]

An early clinical marker for increased risk for hip fracture is the occurrence of a fracture of the distal end of the forearm, (Colles' fracture) in the fourth or fifth decade of life.[1] These patients routinely should have the fact of their increased risk made known to them so that, in discussions with their physician, the benefits and risks of treatment for osteoporosis may be considered.

Medical conditions which are associated with an increased risk for hip fracture include: hemiplegia, or other paresis of a leg, rheumatoid arthritis, previous radiotherapy of the pelvis, diabetes mellitus, and prolonged steroid therapy.[1,29] Among women, low body weight related to body height appears to increase risk.[1]

Most researchers have found the left hip to be involved more frequently than the right.[1,17,20] This finding tends to agree with the observation of greater mineral or ash content in the right femur resulting in increased bone strength on that side.[1]

The benefits of estrogens in protecting against osteoporosis have already been mentioned. More recently, epidemiologic studies have begun to establish the clinical effect of estrogens in protecting against fractures.[30,31] It appears that hip fracture rates may be halved in women under 75 years of age who have taken estrogens for six years or longer.[31] Use for five years or longer seems to impart a more pronounced protective effect than treatment for a shorter duration.[30] Risk for fracture appears to increase after the discontinuation of estrogen therapy, but not to the level of those women who have never been treated with this hormone. The presumed mechanism for such an effect is that bone mass remains higher because of retardation of bone loss during estrogen treatment.[30,31]

Recently, there has been an increasing amount of interest in the potential for fluorides to provide protection from osteoporosis and hip fractures. Fluoridation of water supplies at the level recommended to prevent dental caries (0.7 ppm) does not appear to provide any protection against hip fracture. It is conceivable that higher levels of fluoride could be protective, but that level nor its potential adverse effects are currently known.[32]

MEETING THE CHALLENGE

Currently, there are almost 200,000 hospitalizations yearly for hip fracture among the U.S. population aged 65 and over.[33] We can expect this number to continue to increase over the coming decades, even with advances in preventive care, as the number of very old persons in our society expands.

It is clear that to understand the problem of hip fracture we must also appreciate its antecedents: osteoporosis and falls. The physical therapist can remind the patient with hip fracture to discuss osteoporosis with the physician. The orthopedist frequently is completely concerned with the current problem of the fracture; likewise, the fracture and its precursor may have assumed a lesser importance when the patient again needs routine medical services. Managing osteoporosis is critical to the elderly, and especially to the older white female. Retarding this process can not only prevent fractures, but can also decrease the severity of a fracture and thereby theoretically improve rehabilitation prospects should one occur.[28] We need to learn if falls can be decreased with an exercise program designed to improve balance, build confidence, and increase lower extremity mobility. A randomized controlled clinical trial designed to answer this question should be relatively easy to design and implement. Physical therapists have always advocated the removal of environmental hazards which predispose one to falling. We are also in an excellent position to observe our patients for other clues which may predispose them to this accident, and should report our findings to the attending physician. We are often asked for our opinion whether the patient who has fallen or is at risk for falling should be restrained. Giving our advice could become easier and more precise with the development of a simple statistical formula based on known risk factors for predicting individual risk for falling or falling and fracture.

Finally, in order to determine whether or not our current clinical practice is appropriate for hip fracture treatment and then to assure that we have an adequate supply of therapists who are interested in the problem, we must challenge our educational programs to face the reality of demographic predictions. More vigorous efforts should be made to accept students with an interest in hip fracture and other problems of the elderly and to encourage an interest in geriatrics among all students. We as a profession are also responsible for encouraging our brightest clinicians to obtain the graduate education which will make it possible to answer many of the questions raised in this article. The rapidly expanding geriatric population along with its disproportionate need for health services will otherwise catch us unprepared.

REFERENCE NOTES

1. Alffram PA: An epidemiologic study of cervical and trochanteric fractures of the femur in an urban population. Acta Ortho Scand Suppl No 65:1964

2. The old woman with a broken hip (editorial). Lancet ii: 419–420,1982

3. Wallace WA: The increasing incidence of fractures of the proximal femur: An orthopedic epidemic. Lancet i: 1413–1414, 1983

4. National Center for Health Statistics. Inpatient Utilization of Short-Stay Hospitals by Diagnosis. United States—1974, Series 13-Number 30. U.S. Department of Health, Education and Welfare, Hyattsville, MD, July 1977. DHEW Publication No. (HRA) 77-1783

5. Hielema FJ: The Care and Cure of Hip Fracture in Nursing Homes. Ph.D. dissertation. Chapel Hill, North Carolina, University of North Carolina, 1981

6. Consensus Development Conference. Osteoporosis. JAMA 252: 799–802, 1984

7. Lane JM, Vigorita VJ, Falls M: Osteoporosis: Current diagnosis and treatment. Geriatrics 39: 40–47, 1984

8. Shapiro JR, Rowe DW: Imperfect osteogenesis and osteoporosis. N Engl J Med 310: 1738–1740, 1984

9. Rigotti NA, Nussbaum SR, Herzog DB, et al.: Osteoporosis in women with anorexia nervosa. N Engl J Med 311: 1601–1606, 1984

10. Hulka BS: Effect of exogenous estrogen on postmenopausal women: The epidemiologic evidence. Obstet Gynecol Surv 35: 389–399, 1980

11. Perry BC: Falls among the elderly: A review of the methods and conclusions of epidemiologic studies. J Am Geriatr Soc 30: 367–371, 1982

12. Witte NS: Why the elderly fall. AM J Nurs 79: 1950–1952, 1979

13. Overstall PW: Prevention of falls in the elderly. J Am Geriatr Soc 28: 481–483, 1980

14. Overstall PW: Determining the cause of falls in the elderly. Geriatr Med Today 2: 63–66, 1983

15. Overstall PW, Exton-Smith AN, Imms FJ, et al.: Falls in the elderly related to postural imbalance. Br Med J i: 261–264, 1977

16. Brocklehurst JC, Robertson D, James Groom P: Clinical correlates of sway in old age-sensory modalities. Age Ageing 11: 1–10, 1982

17. Beals RK: Survival following hip fracture: Long follow-up of 607 patients. J. Chronic Dis 25: 235–244, 1972

18. Miller CW: Quality criteria for the treatment of hip fractures. Va Med Mon 102: 1032–1036, 1041–1043, 1975

19. Zimmer JG, Puskin D: An epidemiological model of the natural history of a disease within a multilevel care system. Int J Epid 4: 93–104, 1975

20. Kelsey JL: The epidemiology of diseases of the hip: A review of the literature. Int J Epid 6: 269–280, 1977

21. Fitts WT, Lehr HB, Schor S, et al.: Life expectancy after fracture of the hip. Surg Gynecol Obstet 108: 7–12, 1959

22. Gallannaugh SC, Martin A, Millard PH: Regional survey of femoral neck fractures. Br Med J ii: 1496–1497, 1976

23. Wong PCN: Prevention of femoral neck fractures in the elderly: A Singapore population study. Geriatrics 22: 156–163, 1967

24. Kreutzfeld J, Haim M, Bach E: Hip fracture among the elderly in a mixed urban and rural population. Age Ageing 13: 111–119, 1984

25. Gyepes M, Mellins HZ, Katz I: The low incidence of fracture of the hip in the Negro. JAMA 181: 1073–1074, 1962

26. Iskrant AP: The etiology of fractured hips in females. Am J Public Health 58: 485–490, 1968

27. Miller CW: Survival and ambulation following hip fractures. J Bone Joint Surg (Am) 60: 930–934, 1978

28. Zain Elabdien BS, Olerud S, Karlström G: The influence of age on the morphology of trochanteric fracture. Arch Orthop Trauma Surg 103: 156–161, 1984

29. Pezczynski M: The fractured hip in hemiplegic patients. Geriatrics 12: 687–690, 1957

30. Hutchinson TA, Polansky SM, Feinstein AR: Post-menopausal oestrogens protect against fractures of hip and distal radius: A case-control study. Lancet ii: 705–709, 1979

31. Weiss NS, Ure CL, Ballard JH, et al.: Decreased risk of fractures of the hip and lower forearm with postmenopausal use of estrogen. N Engl J Med 303: 1195–1198, 1980

32. Madans J, Kleinman JC, Cornoni-Huntly J: The relationship between hip fracture and water fluoridation: An analysis of national data. Am J Public Health 73: 296–298, 1983

33. National Center for Health Statistics. Utilization of Short-Stay Hospitals. United States-1982, Series 13-Number 78. U.S. Department of Health and Human Services Hyattsville, MD. August 1984. DHHS Publication No. (PHS) 84-1739

Surgical Treatment
of Arthritis of the Hip

James A. Nunley, M.D.
Eric R. Oser, M.D.

ABSTRACT. Over the last decade, the surgical treatment of arthritis of the hip has undergone a therapeutic revolution, from early days when osteotomies of the femoral shaft or pelvis were the only surgical alternatives to current day techniques which involve partial or complete replacement arthroplasties. The authors provide an overview of the surgical techniques currently available discussing indications and expected results.

INTRODUCTION

The hip joint is an integral part of the locomotive system that transmits load from the body to the ground and, normally, permits bipedal ambulation. The free fluidity with which the normal hip joint functions can be attributed to its anatomy: the hip joint is a ball-and-socket joint and as such is capable of a wide range of motion in various planes. A rounded ball (the femoral head) fits inside a socket (the acetabulum), as Figure 1 shows. This arrangement of acetabulum and femoral head allows rotation in flexion and extension, as well as in abduction and adduction. The intrinsic design of the bony hip joint permits a wide range of motion without any need for strong restraining ligaments, which would restrict such motion. The hip is surrounded by very large muscles that power the body during locomotion; these large muscle groups contribute to the stability of the hip.

Great force can be transmitted across the hip joint. A simple

James A. Nunley, Assistant Professor, Division of Orthopaedic Surgery, Duke University Medical Center, Durham, NC 27710. Eric R. Oser, Hip Fellow, New England Baptist Hospital, Boston, MA 02120.

Reprint requests to James A. Nunley, M.D., Duke University Medical Center, P. O. Box 2919, Durham, NC 27710.

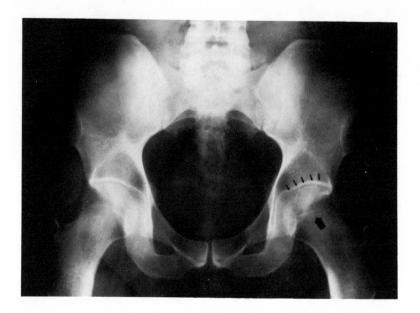

FIGURE 1. AP radiograph of the pelvis demonstrating normal anatomy. Femoral head (large arrow) is spherical and well contained in acetabulum (multiple small arrows). Joint space-area between acetabular bone and femoral head bone is well maintained on both sides.

elevation of the straight leg from the supine position is estimated to produce a force equal to two and one half times the body weight; this force is normally distributed to the articular surface area of the femoral head and acetabulum. Gravitational force increases in the standing posture, and in the one-legged stance phase of normal gait, force on the hip joint increases dramatically, often to several times the body weight.

Any alteration in normal biomechanics of gait will affect one's ability to ambulate painlessly. Arthritis in a hip—whether it is secondary to inflammatory or crystalline disease or degenerative in origin—produces joint-space narrowing and incongruity of the articular surface, with resultant alterations in gait. The first sign or symptom may be a slight abbreviation of the one-legged stance phase of gait. With progression of the disease process and associated joint-space narrowing, smaller areas of the acetabulum must carry greater forces, with the result that more rapid wear occurs in some areas. The ultimate result will be a so-called hip limp or gluteus medius gait. At this point in the disease process, the hip abductors are unable to maintain the pelvis in a level position, and the Trendelenburg sign will become positive. This generally indicates significant hip pathology.

Patients will initially experience pain only with standing or walking. This pain can usually be alleviated at first by rest or weight reduction. As more stress is transmitted to the underlying trabecular bone of the acetabulum, however, some of these trabeculae will hypertrophy and form dense plates of reactive new bone immediately beneath the osteochondral junction. As this hypertrophy occurs, the subchondral plate broadens in an attempt to increase the weight-bearing surface and decrease the load felt by the hip. The result is formation of osteophytes or "spurs," commonly seen on roentgenograms of a painful hip. If stress continues, the unprotected trabeculae in the acetabulum are reabsorbed in localized areas; they then become filled with granulation tissue and necrotic debris and will appear on roentgenograms as cystic areas. The osteophytic response in association with joint-space narrowing is largely responsible for the diminished range of motion common to patients with hip pathology.

If the underlying pathologic process is not interrupted, pain will ultimately be present both day and night, unaltered by weight-bearing or by reclining to rest. An extraordinary osteophytic response may cause autofusion of the hip. If such an arthrodesis occurs, pain will diminish, but the more usual event is drastic reduction of residual motion, with extreme resultant debilitation of the patient.

MODALITIES OF TREATMENT

In an initial evaluation of hip pain, the physician must first confirm the diagnosis. Lumbar spine disease often masquerades as hip pain. Diagnosis of true hip pathology is confirmed through a careful history and thorough physical examination and high-quality roentgenograms of the hip. Once the underlying pathology has been pinpointed, an appropriate course of therapy can then be outlined. Treatment may be nonoperative or operative.

NONOPERATIVE THERAPY

Nonoperative treatment is appropriate for patients with mild disease, minimal pain, very little interference with activities of daily living, and no night pain. Such conservative treatment consists in rest, weight reduction, protected weight-bearing with crutches or a cane, and anti-inflammatory medications, nonsteroidal or, occasionally, steroidal. The agents used and particulars of administration will depend on the patient's age, work status, and the stage of his or her disease.

Surgical treatment of diseased hips is indicated for (1) pain unresponsive to mild analgesics; (2) night pain interfering with sleep patterns; (3) rapidly progressive roentgenographic changes with bone destruction; (4) significant loss of motion, function, or both, interfering with other joints in the lower extremities; and (5) an inability to perform routine daily activities.

OPERATIVE TREATMENT OF HIP PATHOLOGY

Once operative intervention has been decided upon, one of a variety of surgical alternatives must be selected. Major categories of hip surgery are the following: (1) osteotomy of the femur or pelvis, (2) hemiarthroplasty with prosthetic replacement, (3) resurfacing arthroplasty, (4) total joint replacement, and (5) fusion.

OSTEOTOMY

Osteotomy of the femur or pelvis was a common operation in the days before total joint replacement arthroplasty. Osteotomy is suitable for localized disease secondary to increased stress. A successful osteotomy must alter alignment of the bony architecture in such a way as to modify load distribution of the hip. Osteotomy is frequently selected in younger patients with mechanical hip disease resulting from an old slipped capital femoral epiphysis, Legg-Calvé-Perthes' disease, or congenital hip dysplasia. X-ray examination will often reveal narrowing of the joint-space, but only in selected areas, such as the superior weight-bearing surface. Preoperative roentgenograms with the leg in abduction, adduction, and internal and external rotation will demonstrate that the joint space widens in a specific anatomic position, and this is generally the position selected for osteotomy (Figure 2). The initial deformity is most commonly seen within the femoral head, thus requiring an osteotomy of the femur. This will usually be a valgus or varus osteotomy, but flexion/extension or rotation osteotomy may occasionally be performed.

In a painful hip with congenital subluxation or dislocation, the pelvis is usually deficient and is thus selected as the site of osteotomy. A Chiari osteotomy—transection of the innominate bone at the level of the hip joint—will provide superior and lateral coverage for the femoral head, helping to redistribute the forces (Figure 3). In most cases, osteotomy provides years of pain-free living.

HEMIARTHROPLASTY

Hemiarthroplasty, the replacement of one half of the hip joint, is generally referred to as replacement of the femoral head and neck. This operation is usually reserved for younger patients whose disease is limited to the femoral head—in most cases, aseptic necrosis. Initially, changes are present only on the femoral side of the joint, but after years of symptoms, the acetabulum becomes deformed and arthritic. Hemiarthroplasty is appropriate only for patients who have normal or near normal acetabula at the time of surgery. There are many types of hemiarthroplasty. One of the earliest was replacement of the femoral head with the A. T. Moore prosthesis. This prosthesis was initially used for elderly patients with hip fractures, but its excellent design is also suited to use in younger patients with hip disease. The A. T. Moore prosthesis is a solid metallic sphere connected to a long metallic stem, which is inserted in the shaft of the femur (Figure 4). Its disadvantage is that with the passage of time, the acetabulum wears down from contact with the metallic sphere. A newer type of hemiarthroplasty, designed to overcome this problem of acetabular deterioration, is the bipolar prosthesis (Figure 5). In the bipolar prosthesis, the metallic replacement for the femoral head is covered by a polyethylene cup, which is enclosed in a metal bearing. In such a prosthesis, friction occurs between the plastic end bearing and the metal acetabulum and *not* between the metal outer covering and the

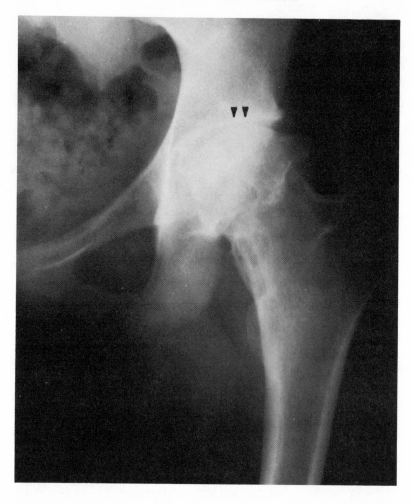

FIGURE 2. a. Sixteen year old patient with healed Legg-Perthes' Disease. Femoral head irregularly shaped and superior weight-bearing dome (arrow) shows moderate joint space narrowing.

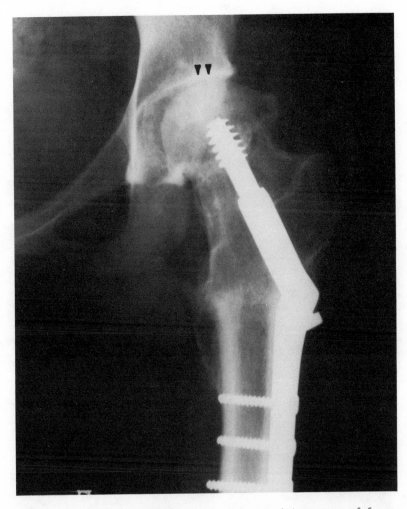

FIGURE 2. b. Postoperative appearance of valgus extension osteotomy of femur. Osteotomy performed in subtrochanteric region and internally fixed with sliding screw device. Note increased joint space (arrow) between femoral head and acetabulum when compared with preoperative appearance.

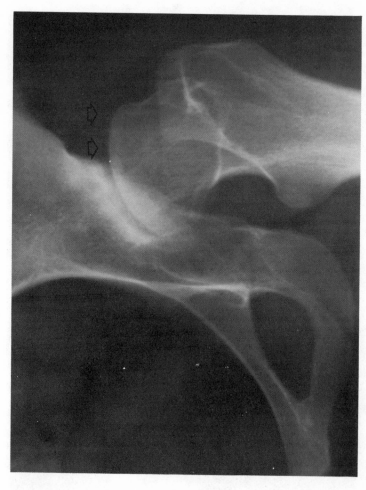

FIGURE 3. a. AP radiograph of hip preoperatively in 31 year old female with old congenital dysplastic hip. Note the marked uncoverage of the femoral head laterally (arrow).

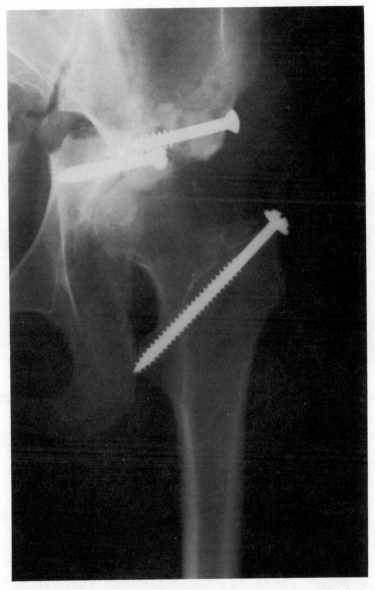

FIGURE 3. b. Postoperative radiograph of Chiari pelvic osteotomy. Note medial displacement of the pelvis with lateral coverage of femoral head. Large metallic screws fix osteotomy. Screw with washer in greater trochanter was to secure trochanteric osteotomy used for exposure only.

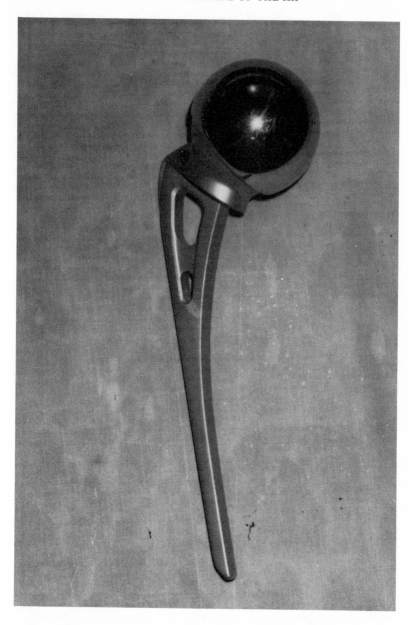

FIGURE 4. A. T. Moore prosthesis with curved stem and fenestrations for bone ingrowth.

original bony acetabulum. This biopolar prosthesis should spare the acetabular cartilage and generally provide years of painless motion.

RESURFACING ARTHROPLASTY

Resurfacing arthroplasty was initially carried out, as the name implies, by resurfacing one or both joint surfaces with a synthetic material. One of the earliest types was the cup arthroplasty, in which a small portion of the femoral head was removed and a metallic cup was inserted over the remaining femoral head and neck. In many patients, this operation provided good temporary relief, but inevitably the bone beneath the cup deteriorated or the acetabulum wore down, as in the hemiarthroplasty patients.

The next major advancement in the field of resurfacing arthroplasties was conceived of as an alternative to total joint replacement for younger patients in whom one wished to preserve bone stock. In this type of resurfacing arthroplasty, a very small portion, perhaps 5 mm, of bone is removed from the acetabulum and a polyethylene cup is cemented into the acetabulum; next, several millimeters of bone are removed from the femoral head, and a metallic cup is placed over the bone of the femoral head and cemented into place (Figure 6). In the early 1980's, resurfacing arthroplasty was very highly thought of, but recent results seem to indicate that the problems of loosening and eventual failure of the prosthesis are significant. At present there seems to be little justification for the resurfacing arthroplasty.

TOTAL HIP ARTHROPLASTY

Total hip replacement entails replacement of the femoral head and neck and the acetabulum by synthetic components. Total hip replacement was initially performed like a hemiarthroplasty—that is, the femoral stem and acetabulum were *press* fitted into bone. With repetitive motion, however, these components loosened and became so painful as to warrant reoperation. Sir John Charnley's contribution ultimately made total hip replacement a universally performed operation with predictably excellent results. Charnley conceived the idea of cementing the two components of the arthroplasty in place, using polymethylmethacrylate to bond bone to

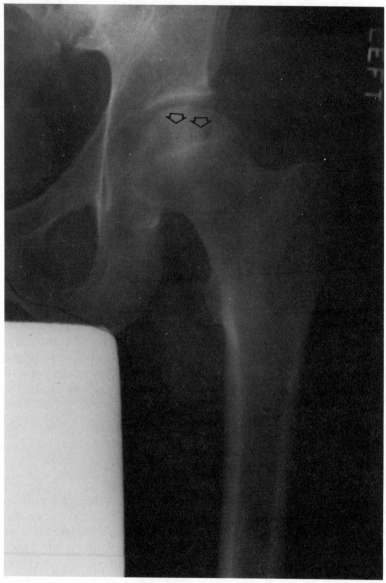

FIGURE 5. a. AP radiograph in 48 year old male with painful left hip. Central area of femoral head demonstrates fracture through area of avascular necrosis (arrow). Joint space well maintained and acetabulum congruous.

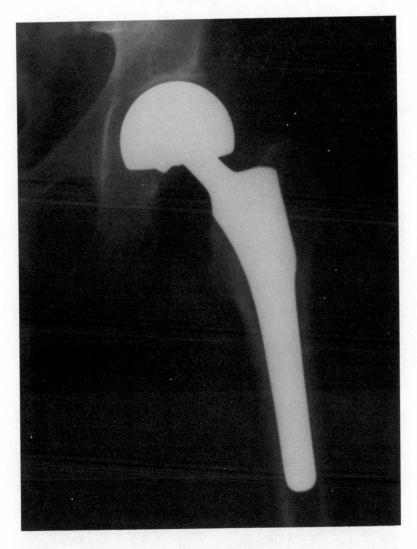

FIGURE 5. b. Postoperative radiograph of bipolar hemi-replacement arthroplasty. Femoral prosthesis includes small metallic head covered by polyethylene cup and finally covered with large metallic sphere. Motion will occur between metal and polyethylene cup rather than between metal and bone of acetabulum.

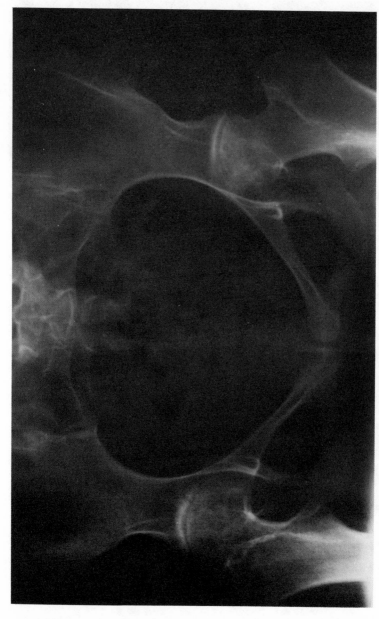

FIGURE 6. a. AP pelvis radiograph preoperatively in 32 year old female with bilateral rheumatoid arthritic changes of hip.

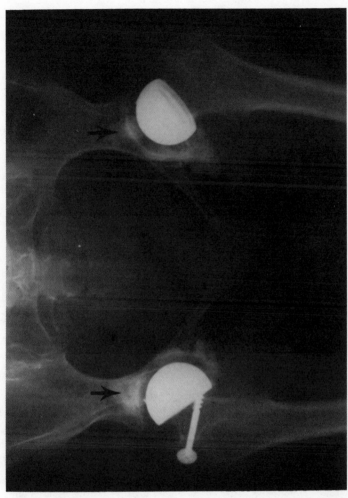

FIGURE 6. b. Postoperative AP radiograph of bilateral resurfacing arthroplasties. On one side trochanteric osteotomy was performed for exposure. Note screw transfixing trochanter back to femur. Acetabulum secured to pelvis with cement (arrow) and femoral head is covered by the metallic sphere.

prosthesis. This cementing heralded a new era in hip surgery, making possible an excellent operative procedure that nearly always yields a painless hip with excellent biomechanics. Such a result is gratifying for both patient and physician (Figure 7). Problems encountered in long-term studies of total hip arthroplasty have centered around the loosening which occurs at the cement-bone interface. When loosening occurs, the hip once again becomes painful. The bone may subside beneath the prosthesis, and an exchange of hardware and new cement will usually be necessary about ten years after the initial arthroplasty.

These long term problems with cemented prostheses provided the impetus for development of the porous-coated total hip replacement arthroplasty. In the porous-coated arthroplasty, the hip joint components both femoral and acetabular prostheses are lined with a substance that allows bony ingrowth into the prosthesis. The metallic devices are thus fixed to the patient's own bone by the formation of bony trabeculae within the prosthesis (Figure 8), obviating the need for cement. The anticipated benefit is prevention or diminution of future loosening after arthroplasty. This procedure is used primarily in younger patients; early results have been encouraging, but it is too early to predict the long-term results of such an operation.

HIP FUSION

For the young, vigorous, working patient with unilateral hip pathology, the best solution for a painful hip may be hip fusion (Figure 9). Elimination of motion in the hip joint should provide the patient with a painless extremity, allowing him to work under heavy load with a good gait and even to do heavy manual labor. In contrast to patients who undergo total hip arthroplasty, there are no restrictions on one's activity after a successful hip fusion. There are, however, some general problems associated with arthrodesis of the hip joint.

Because of the altered biomechanics of gait, flexion and extension of the lower extremity must be performed through the lower lumbar spine. The result is, frequently, advanced degenerative arthritis of the lumbar spine and, after many years of altered gait, severe incapacitating back pain. At this point, the surgeon may elect to take down the hip fusion and convert it to a total replacement arthroplasty.

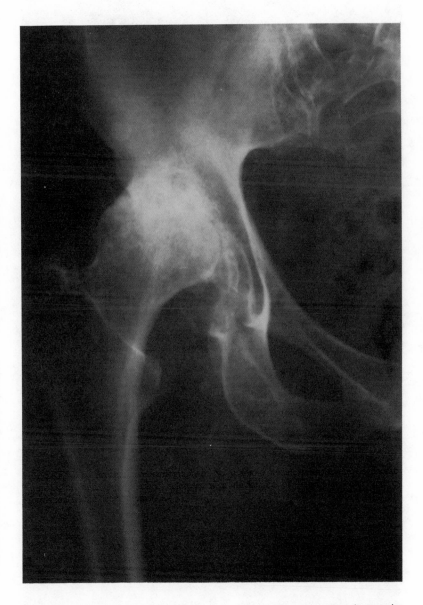

FIGURE 7. a. Preoperative radiograph of 86 year old female with severe degenerative arthritis of right hip. Marked sclerosis with thickening of trabecular bone and cyst formation within acetabulum.

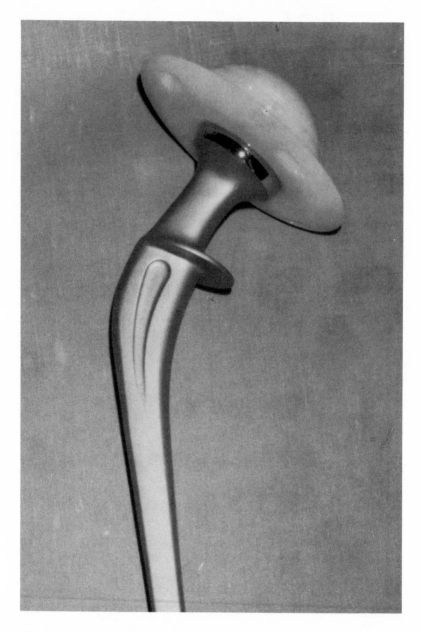

FIGURE 7. b. Typical total hip replacement arthroplasty components. Metallic femoral stem with collar for insertion into femoral shaft and polyethylene acetabular cup which is secured to pelvis with cement.

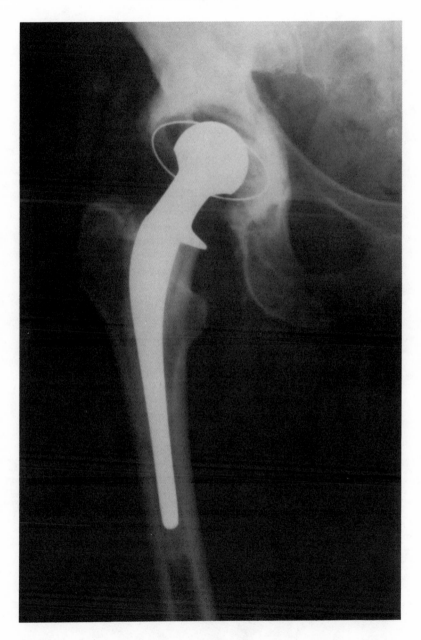

FIGURE 7. c. Postoperative radiograph of total hip replacement arthroplasty well inserted into bone of femur and acetabulum.

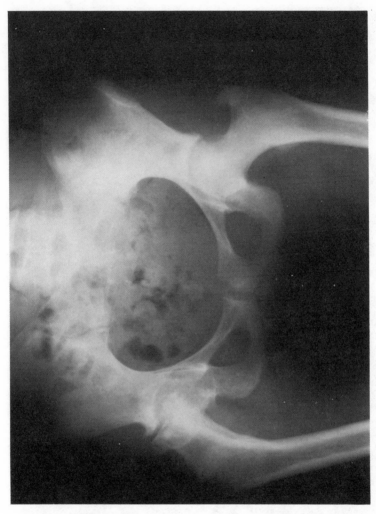

FIGURE 8. a. AP radiograph of 24 year old female with marked irregularity to femoral head and acetabulum.

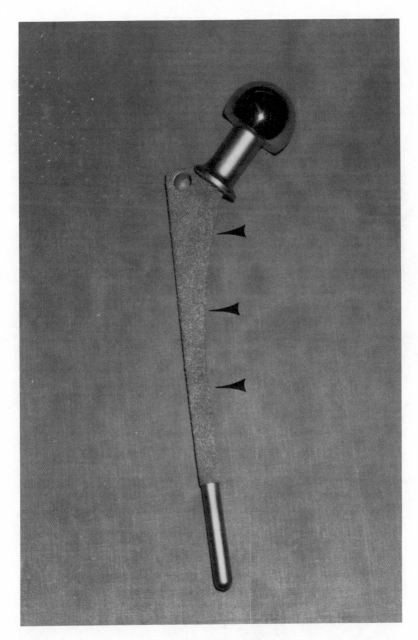

FIGURE 8. b. Photograph of a porous coated prosthetic femoral stem. Greyish modeled area (arrows) indicates porous coating which will allow bony ingrowth into prosthesis. Stem and head of prosthesis are smooth metallic surfaces.

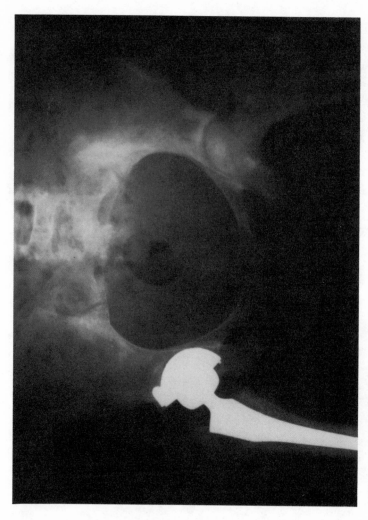

FIGURE 8. c. Postoperative radiograph of patient in (a) demonstrating porous coated acetabular and femoral components of total hip arthroplasty.

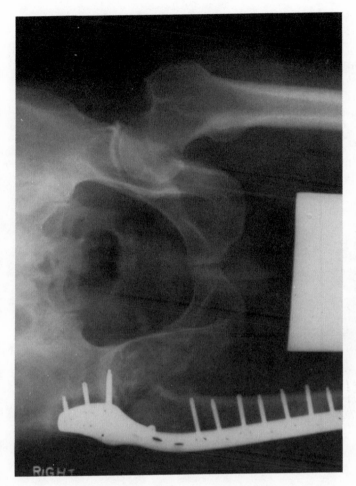

FIGURE 9. Postoperative radiograph of patient undergoing hip fusion. Note absence of femoral head. Bone graft inserted into acetabulum and into femoral canal. Femur fixed to pelvis with large metallic plate. Six screws proximally into pelvis and eight screws distally into femur secure these two bones together.

SURGICAL TECHNIQUE

One of three general surgical approaches will serve any operative procedure on the hip. These are the anterolateral, lateral, and posterior approaches. Each has advantages and disadvantages.

The anterolateral approach is made in the interval between the tensor fascia lata and gluteus medius muscles. This approach allows anterior exposure of the hip joint, and a complete anterior capsulectomy of the hip may be performed by using it. This approach can also be used for an osteotomy.

The direct lateral approach is made either by means of a trochanteric osteotomy or by a gluteus medius muscle-splitting incision. Its advantage is in wide exposure of the anterior and posterior rims of the acetabulum. Total hip arthroplasty technically is easier through a lateral exposure, but trochanteric osteotomy in combination with total hip arthroplasty carries a risk of nonunion. Should the osteotomy of the greater trochanter fail to heal, there is often pain, which compromises the results, and also there may be instability leading to dislocation.

The direct posterior approach is a muscle-splitting incision through the gluteus maximus. This approach entails detachment of the short external rotators of the hip as well as posterior capsulectomy. It is very commonly used and allows excellent exposure of the acetabulum. The difficulty with this approach is the accompanying frequency of posterior dislocation of the hip in the perioperative period.

CONCLUSION

Many great advances have been made in the care and surgical treatment of arthritis of the hip in the last twenty years. There are many new technical developments on the horizon, and the future of surgical treatment of hip pathology appears bright.

Survey of Physical Therapy Preoperative Care in Total Hip Replacement

Patricia Bickerstaff Ball, M.Ed., P.T.
Martha C.Wroe, M.A., P.T.
Loretta MacLeod, P.T.

ABSTRACT. The purpose of the clinical survey is to determine the current preoperative evaluation procedures and instructions given to patients receiving Total Hip Replacement (THR) surgery. Sixty-two Florida hospitals were surveyed with a sixty-one percent of survey return. The most frequent assessment procedures currently being used and subjective benefits of preoperative care and instructions for this patient population are included in this report.

In 1984, approximately 100,000 total hip replacement (THR) surgeries were performed in the United States.[1] Although physical therapists have been treating this patient population since the advent of the procedure in the 1960's, little information has appeared in the recent physical therapy literature explaining the current treatment for this group of patients. The purpose of this inquiry was to survey a number of physical therapy departments in Florida hospitals where THR surgeries are performed and to gather information concerning current preoperative evaluations and instructions given to patients.

During the 1970's several preoperative protocols were presented in the physical therapy literature. In 1972 Beber and Conveny reviewed the preoperative, postoperative and outpatient physical therapy management of patients with Ring, Charnley, and

Patricia Bickerstaff Ball, Director of Physical Therapy, Department of Rehabilitation Medicine, GVAMC, Gainesville, FL 32602.
Martha C. Wroe, Professor, Department of Physical Therapy, JHMHC, Gainesville, FL 32610.
Loretta MacLeod, ATC, Staff Physical Therapist, Shands Hospital, Gainesville, FL 32610.

Charnley-Müller total hip prostheses. The preoperative evaluation developed by William Harris was used. This form assessed the hip with a point rating system for pain, functional ability and motion with 100 points possible for a normal hip.[2]

Thieler and Mueller included in their preoperative instructions breathing exercises and orientation to a spring and sling suspension that was used several days postoperatively to ensure hip stability.[3]

Schamerloh and Ritter reviewed 502 total hip replacement surgeries and described their preoperative program to include evaluation of muscle strength, joint ROM and ADL level. Patients were instructed in postoperative routines which included THR precautions.[4]

In 1975 Richardson reported on the management of 327 patients with hip arthritis who underwent THR arthroplasty between 1970–1974. The preoperative evaluation became part of the patient's continuing data base. Included were baseline tests for ambulation and transfer ability, ROM, strength, sensation, orientation of patients, psychosocial information and instructions in exercise, transfers, and ambulation methods. The preoperative exercise program included static exercises for hip extensors, quadriceps and hip abductors, and active exercises for ankles and upper extremities, breathing exercises, appropriate transfer techniques and proper bed positioning. Ambulation aids were also chosen and patients received appropriate gait training. The author commented that a successful preoperative physical therapy program was contingent on the following:

1. A system which assures 100% patient referral
2. Completion of the evaluation within a reasonable time frame
3. Recognition that strength testing results have limited value when significant pain is present
4. Communication of results to the patient's physician
5. Documentation of results
6. Collaboration/communication with other appropriate allied health personnel
7. Initiation of discharge planning during the preoperative evaluation; family members instructed in exercises and THR precautions[5]

METHOD

Sixty-two hospitals in Florida were surveyed regarding physical therapy preoperative care of patients receiving THR surgery. The questionnaire used is illustrated in Figure 1.

RESULTS

Thirty-eight (61%) of the surveys were returned, but all questions were not completed by the participants.

The number of patients receiving THR surgery treated annually ranged from less than 20 to approximately 600 patients. The mode was 75 to 100 patients yearly.

Preoperative instructions were given in 17 hospitals compared to 16 who did not give preoperative treatment. Three respondents indicated preoperative care had been discontinued since the initiation of diagnoses related groupings (DRGs). Four physical therapists responded that some referring physicians within their hospital did not request preoperative instructions. One facility was a trauma center, and preoperative instructions were not possible. Sixteen hospitals used a standard preoperative assessment in contrast to 22 who did not use a special form.

Frequency of assessment procedures are demonstrated in Figure 2.

The most frequent preoperative procedures used were observational gait analysis (gross), isometric exercises, instructions in THR precautions and gait training. Transcutaneous electrical nerve stimulation (TENS) instructions were given preoperatively in three facilities.

Two departments reported collection of data relative to the benefits of preoperative instructions. The method of study and criteria established were not explained, but the benefits cited were reduction of anxiety, faster and smoother postoperative progression, and more realistic patient goals. Additional subjective benefits listed by other facilities included better patient follow through with exercise programs and increased patient comfort. One facility reported decreased rehabilitation time. Another respondent stated that preoperative care was of minimal benefit to the patient but helpful to the physical therapist in assessing the abilities of the patient. The most frequent benefits mentioned were a reduction in patient anxiety and increased patient comfort.

SURVEY OF PHYSICAL THERAPY PRE-OPERATIVE CARE

FOR TOTAL HIP REPLACEMENT PATIENTS

1. Please estimate the number of THR patients seen in your department annually.

　　　　　　　　　　　　　　　　　　　　　　　　　　　_____

2. Are pre-operative orders routinely sent to Physical Therapy?

　　　　　　　　　　　　　　　　　　　　　　　　YES　　　NO

　　　　　　　　　　　　　　　　　　　　　　　　____　　____

3. Does your department use a standard pre-operative form?

　　　　　　　　　　　　　　　　　　　　　　　　YES　　　NO

　　　　　　　　　　　　　　　　　　　　　　　　____　　____

4. If not, please indicate why your department prefers not to use a standard form.

　　_____

　　_____

　　_____

5. If a standard form is used, please indicate which of the following are included in the THR pre-operative evaluation/instruction program.

		YES	NO
a.	ROM measurements	____	____
b.	Strength evaluation	____	____
c.	Sensory assessment	____	____
d.	Assessment of pain	____	____
e.	Gross gait evaluation	____	____
f.	Gross ADL evaluation	____	____
g.	Instruction in breathing exercises	____	____
h.	Instruction in isometric exercises	____	____
i.	Instruction in active exercises	____	____
j.	Review of THR precautions	____	____
k.	Instruction in gait training	____	____

6. Has your department done follow-up collection of data to assess changes in motivation or other influences of THR pre-operative instructions?

　　　　　　　　　　　　　　　　　　　　　　　　YES　　　NO

　　　　　　　　　　　　　　　　　　　　　　　　____　　____

7. Please report briefly your clinical observations of benefits obtained from pre-operative THR instructions to patients.

　　_____

　　_____

　　_____

Thank you for your assistance. Please return no later than March 15, 1985.

FIGURE 1. Survey Form

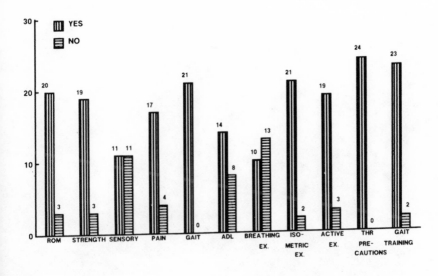

FIGURE 2. Preoperative Assessment and Treatment Procedures

SUMMARY

There have been few studies in recent physical therapy literature exploring current treatment of patients receiving THR surgery. The purpose of this inquiry was to survey a number of physical therapists regarding the current preoperative evaluation and instructions given to this patient population. Although great variability was reflected in the survey, many respondents cited subjective benefits. Future studies might clarify the exact subjective as well as possible objective benefits of physical therapy preoperative instruction in patients receiving THR surgery.

REFERENCES

1. In a communication with William Petty, M.D., Professor and Chairman, Department of Orthopedics, University of Florida, Gainesville in April, 1985
2. Beber CA, Conveny R: Management of patients with total hip replacement. Phys Ther 52:823–828, 1972
3. Thieler PC, Mueller KH: Immediate postoperative management of patients with total hip replacement. Phys Ther 53:949–955, 1973
4. Schamerloh BS, Ritter MA: Prevention of dislocation or subluxation of total hip replacements. Phys Ther 57: 1028–1031, 1977
5. Richardson R: Physical therapy management of patients undergoing total hip replacement. Phys Ther 55:984–989, 1975

Physical Therapy for Patients
with Hip Fracture or Joint Replacement

Frank Hielema, Ph.D., P.T.
Rebecca Summerford, L.P.T.

ABSTRACT. The purpose of this paper is to review the role of physical therapy in hip fracture and hip arthroplasty in the adult hip. The first portion is devoted to the hip fracture patient including screening for hip fractures, risk factors for nonrecovery, and treatment guidelines for each of the approaches used for fracture stabilization. The second portion is devoted to the hip arthroplasty patient, both hemiarthroplasties and total hip replacements (THR), including surgical approach, recovery, rehabilitation, and postoperative complications.

The two primary conditions of the adult hip necessitating referral to physical therapy are hip fracture and hip arthroplasty. While the goals of treatment for both conditions are similar, the populations experiencing hip fracture and hip arthroplasty are quite different. Hip fracture repair is an emergency procedure which befalls a cross-section of the elderly population. In fact, many of these individuals who have become prone to fall represent the frailer segment of this population. On the other hand, persons *chosen* for hip arthroplasty represent a *select* segment of the adult population (usually elderly) who are good surgical risks and are expected to derive benefit from the procedure. Consequently, by the very fact that individuals selected for hip arthroplasty are generally healthier

Frank Hielema, Director of Rehabilitation and Research, Hillhaven Convalescent Center, and Adjunct Instructor, Department of Epidemiology, School of Public Health, University of North Carolina, Chapel Hill, NC 27514, and Chairman, North Carolina Governor's Council on Physical Fitness and Health.

Rebecca Summerford, Staff Physical Therapist, Hillhaven Convalescent Center, 1602 East Franklin Street, Chapel Hill, NC 27514.

The authors wish to thank Brigitte Scek, Wray Fleming Powell, and Donna Aliyetti for their assistance in the preparation of this manuscript, and Edwin T. Preston, M.D. for his thoughtful review of content.

than those fracturing a hip, we can expect higher achievements from those undergoing this elective procedure. While we struggle frequently to maintain life or even approximate the level of preinjury function among our hip fracture patients, clients with total hip arthroplasties amaze themselves and sometimes even the health care team with tales of regaining skills lost months or years previously.

This paper will discuss the role of physical therapy in hip fracture and prosthetic hip replacement. The first part is a review of screening procedures that therapists may use to alert physicians to whether or not a fracture has been suffered. Risk factors for nonrecovery from a hip fracture will be discussed. Treatment guidelines for the various approaches used to stabilize hip fractures will be presented.

The second portion will include a brief description of the surgical approach used for hemiarthroplasties and total hip replacements (THR) coupled with information concerning the recovery of patients undergoing such surgical procedures. Both types of hemiarthroplasties, cemented and porous-coated, along with THR will be discussed in detail with respect to rehabilitation and postoperative complications from the point of view of the physical therapist.

HIP FRACTURE

The Role of the Physical Therapist in Screening for Hip Fracture

Physical therapists, particularly those practicing in long-term care settings, may be called upon to offer an opinion as to whether or not a patient has suffered a hip fracture in a fall. Such options serve as a screening procedure which, when operating at its best, would send all fracture cases (true positives) to the radiologist and orthopedist while avoiding the expense and trouble of x-raying those who have not suffered a fracture (true negatives). According to the sensitivity and specificity of the screening procedure, a certain number of fractures will not be identified (false negatives) and some patients who have not broken a hip will be sent on for further evaluation (false positives).[1]

Examining for a triad of symptoms frequently is helpful in identifying patients who have fallen who should receive medical evaluation. If the leg suspected of being fractured is lying in external rotation, is shortened, and is painful with attempts at active or passive movement, examination by the patient's physician is indicated. A technique reported to be both more sensitive and specific has recently been described in screening for hip fractures.[2,3] A stethoscope is placed over the symphysis pubis and then each patella is percussed. If a fracture of either the hip or the pelvis is present, the sound reaching the examiner's ear will be lower in pitch and volume on the affected side than on the uninvolved side. While Carter has described a discontinuity in sound conduction in impacted fractures,[2] Berger's experience has shown that the test may produce a false negative result when impaction of the fracture is present.[3]

Whenever the physical therapist has any suspicion that a hip fracture may be present, evaluation by the physician is essential. Both medically and legally, the therapist should err on the side of being overly cautious.

Risk Factors for Nonrecovery From Hip Fracture

The patient with hip fracture almost always enters into the health care system, usually by being admitted directly into an acute care hospital.[4,5] Only the frailest, who are such high surgical risks that surgery cannot be considered, will remain at their home or in a nursing home. The goal of treatment is to allow as rapid mobilization as possible toward achieving independence in ambulation and activities of daily living while not jeopardizing chances for bone healing.[6] Just over 50 years ago a tremendous acceleration in rehabilitation was achieved with the introduction of internal fixation allowing early mobilization. This prevents the complications of prolonged bed rest[7,8] which otherwise would occur if the fracture were allowed to heal without surgical repair.

Researchers have begun to generate a considerable body of knowledge related to factors which influence outcome following hip fracture. Some studies have dichotomized outcome into life and death,[4,9–12] while others also have considered function among survivors.[5,13–17]

Persons suffering a hip fracture are at greater risk for death than those in an uninjured age, sex, and race comparable population. This increased risk has been found to end by three months,[4,18] six months,[9] eight months,[17] or one year post-fracture.[10,11] One year survival rates vary from 50 to 78 percent.

Advanced age and a high prevalence of secondary or complicating diagnoses consistently have proved to increase both the probability of death or decreased functional status following hip fracture.[4,5,9–11,15–17,19] The value of age as a predictive factor for a poor outcome may decrease after the influence of complicating diagnoses is controlled.[15] The presence of dementia has been found to be especially devastating to the chances of recovery from hip fracture, and the risk for nonrecovery appears to increase as the degree of cerebral dysfunction increases.[15] Other complicating diagnoses which have been found by more than one investigator to impart a higher risk for poor outcome include Parkinson's disease, active cancer, and renal failure. The presence of arteriosclerotic heart disease, diabetes, excessive obesity, rheumatoid arthritis, multiple sclerosis, surgical wound sepsis, and pressure ulceration have each been found, in at least one study, to be predictive of poor outcome. Male gender and trochanteric fracture type both have been found to be associated with increased risk for poor outcome, but it is likely that the associations are secondary. The relationship between improved chances of survival and more severe trauma as a cause of the fracture appears to be the result of younger, healthier individuals being those who tend to put themselves at risk for moderate to severe trauma.[4] No form of anticoagulant therapy convincingly has been shown to have a significant effect on total mortality in hip fracture cohorts. Only when venous thrombosis or death from pulmonary embolus has been the outcome of interest has a significant benefit from oral anticoagulation been demonstrated.[20–25]

Repair and Rehabilitation After Hip Fracture

The rehabilitation program following hip fracture and its rate of progression to a large extent are determined by the treatment used to stabilize the fracture and the degree of stability achieved. The treatment is based largely on the characteristics of the fracture; therefore, a brief review of the classification of hip fractures is in order.

The hip joint is surrounded by an articular capsule. Proximally, the capsule is attached in the region of the acetabular labrum; its distal or femoral attachment is along the intertrochanteric line anteriorly, medially just above the lesser trochanter, and posteriorly to the neck of the femur.[26,27] Whether the line of a hip fracture primarily is included outside or within the joint capsule distinguishes whether the fracture is classified as extracapsular or intracapsular. Extracapsular fractures include intertrochanteric (between the lesser and greater trochanters) and pertrochanteric (through the trochanters) fractures. Intracapsular fractures also may be referred to as femoral neck fractures. The location of the break within the capsule may be refined with classifications of subcapital, transcervical (midcervical), or basilar (base of the neck).[28] The degree of displacement present[29] in an intracapsular fracture is of special importance because of the potential for interruption of the arteries supplying the head of the femur.[26,28]

The vast majority of older individuals suffering a hip fracture will receive surgical fixation although treatment in some cases may rely more conservatively on casting, traction,[30,31] or rigid air splinting.[32] The goal of internal fixation is to provide sufficiently rigid internal stabilization of the fracture fragments to allow mobilization without the need for external support. Rapid mobilization promotes full function of the joints and muscles while avoiding the complications of immobility, namely: muscle atrophy, joint contracture, disuse osteoporosis, and edema.[28] The physical therapist while promoting rapid mobilization has a responsibility to insure maintenance of the internal stabilization achieved. All the king's horses and all the king's men may not be able to achieve a sufficiently rigid fixation to allow rapid resumption of full function of an obese woman in whom a comminuted intertrochanteric fracture or intracapsular fracture of the hip has been fixed with Knowles' pins. Prosthetic replacement may allow more rapid return to full weight-bearing and function, but this treatment carries added risks of infection or prosthesis dislocation.[28]

Among patients treated with casting or traction, physical therapy is limited initially to instructions for exercising the sound leg and providing maximal independent bed mobility. When traction is the chosen treatment, range-of-motion exercises for the knee joint of the injured limb may begin after several weeks of restricted mobility has allowed bone union to begin.[31] Individuals with casted fractures should be given maneuvering and safety instructions for a wheelchair, as soon as this progression is allowed, in order to provide freedom of movement during the convalescent phase. Treatment with a rigid air splint reportedly allows immediate mobilization with full weight bearing on the fractured hip.[32]

A detailed discussion of the rationale for choosing any of the various surgical treatments for hip fracture is beyond the scope of this paper. Some mention, however, should be made of the general effect that the choice of surgical hardware has on the rate of mobilization of the patient.

Extracapsular fractures commonly are treated with a combination of a hip nail which crosses the fracture line connected to a side plate with screws firmly attaching the plate to the distal fracture segment. Usually, the patient progresses from non-weight-bearing to touch-down or partial weight-bearing in the first few weeks postoperatively. Normally, return to full weight bearing is delayed until union of the fracture site has occurred unless the orthopedist feels that there is only a small risk for metal failure with early full weight-bearing. For example, risk for failure might be relatively small in the slightly built, or relatively inactive, patient with a stable fixation. Conversely, unstable comminuted fractures may safely tolerate no more than non-weight-bearing or touch-down weight-bearing status for the first few weeks or months.

Less commonly, extracapsular fractures may be fixed with flexible intramedullay Ender's nails inserted into the femur at its medial distal end and guided across the fracture site.[6,28,33] While Ender's nailing involves minimal surgical trauma[6] and allows for a rapid return to full weight bearing, anatomical fracture reduction is more difficult to achieve and maintain.[6,34]

Intracapsular fractures are commonly treated with a combination nail and side plate, multiple pins (e.g., Knowles' pins), or prosthetic head replacement. Multiple pins are more likely to be chosen for nondisplaced or impacted femoral cervical fractures.[28,35] Percutaneous insertion of multiple pins is a relatively simple method of internal fixation; however, weight-bearing must usually be delayed to avoid placing greater stresses on the narrow diameter pins than they can accommodate. If full weight-bearing is resumed before fracture union occurs, the pins may actually break.[35] Femoral head replacement generally is chosen for treatment when the vascular supply to the femoral head is suspected to have been interrupted, if satisfactory closed reduction cannot be obtained, or if the fracture has resulted from skeletal metastasis of cancer.[28] While femoral head replacement with a cemented prosthesis usually allows a rapid return to weight-bearing, risks include serious infection, dislocation during convalescence, and, later, prosthesis or cement failure. More attention will be devoted to prosthetic head replacement in the section of this paper dealing with hip joint replacement.

The physical therapy evaluation of the postsurgical hip fracture patient should begin with knowledge of the type of fracture fixation used, and the orders for weight-bearing. The therapist must be aware of any complicating medical conditions and of any potential these conditions have on limiting rehabilitation. Determining the individual's preinjury functional status and living arrangements, as well as the availability of social support is important for establishing short and long range goals. Assessment of the patient's prefracture as well as current mental status, orientation, and ability to follow instructions is essential for predicting the long range value of physical therapy in the convalescent period. Even in the moderately demented patient, however, a trial of physical therapy is often warranted to determine if any benefit can be gained. A gross evaluation of the strength of the upper extremities and the unaffected lower extremity is helpful. While manual muscle testing about the affected hip is contraindicated, assessment of the strength of the muscle groups acting below the hip on the affected side is appropriate. Range of motion of all joints of the affected limb should be evaluated. Precautions must be taken to keep hip motion within the limits of preventing prosthesis dislocation when this form of repair has been used. These limits are discussed in the following section on hip arthroplasty. Functional evaluation should include the patient's ability to change his position in bed, particularly the position of the affected limb, his ability to go from supine to sitting and return to supine, his ability to come to standing from a seated position and then to transfer. Determining an individual's ability to operate a wheelchair safely will indicate whether he is ready to or will need instructions to use the chair as a temporary means of mobility. The evaluation should include the patient's ambulation status specifying the assistive device used, distance, gait quality, and ability to maintain weight-bearing precautions and balance. Throughout the evaluation, the therapist continuously should be aware of the patient's pain level in the affected limb. The wise therapist will inquire about the circumstances surrounding the patient's fall. Was the fall due to an environmental obstacle; and if so, can steps be taken to make the environment less hazardous?[36] Or, is it suspected that the fracture led to the fall? Finally, has the patient discussed osteoporosis with the physician?[15,37]

A specific treatment plan with short and long range goals is then developed to overcome the individual's deficits. If the nursing staff has any reservations about technique in positioning or transferring the patient, the therapist can review the instructions for the particular patient. For example, if a knee immobilizer is being worn on the affected limb to prevent excessive hip flexion when femoral head arthroplasty has been the treatment of choice for the hip fracture, reminders to relieve pressure over the head of the fibula in application of the immobilizer may help to avoid peroneal nerve palsy. Gait training is initiated, and the patient is encouraged to adhere to the weight bearing limits established by the orthopedist. Endurance, gait quality, coming to standing, and safe return to a chair are emphasized. If indicated, instruction in wheelchair maneuvering should be included in the treatment program. A method for safe, independent movement onto the commode either using the walker or a transfer from the wheelchair should be developed. Active-assistive range-of-motion exercises are given as soon as possible to the affected limb within the limits allowed by the surgical procedure.[38,39] The patient may also be taught isometric contractions of the quadriceps and gluteals if this maneuver is understood. Progression to active motion of the involved extremity is encouraged over the next several weeks,[38] with emphasis placed on the quadriceps, hamstring, and gluteal muscle groups.[40] Strengthening exercises are indicated for the unaffected extremities if weakness has been found in the evaluation.

Resisted strengthening of knee extension in a short arc with weights up to ten pounds and with a bolster under the knee usually poses no problems in patients with hip fracture. Straight leg raises, bridging, and gravity resisted hip abduction exercises, however, are absolutely contraindicated until bony union is established in all patients with hip fracture except those with a firmly cemented prosthesis without trochanteric osteotomy. In these exercises, the long lever arm of the leg distal to the fracture site places excessive strain on the fracture site and the fixation device, which may contribute to device failure. Additionally, such exercises can contribute to continued movement at the fracture site if rigid internal fixation has not been achieved, resulting in nonunion of the fracture.[28] Plans must be made for training in stair climbing if this skill is necessary for return home.

Frequently, in his convalescence, the patient with hip fracture may move through several levels of care such as progression from hospital, to skilled nursing facility, to home. Communication among therapists in all of these settings will help to insure that the patient's program will be continued at the appropriate level of activity with each successive placement and that his individual goals will be attained.

Some discussion of the special case of the demented patient with hip fracture is in order. It has been pointed out previously that rehabilitation potential is decreased in the presence of dementia, and that the risk for nonrecovery increases in direct proportion to the degree of cerebral dysfunction. It is important for the physical therapist to appreciate whether dementia was present prior to the fracture, and whether the injury and removal from familiar surroundings have exacerbated the cognitive disability. Frequently, with establishment of a routine in physical therapy, and with frequent reinstruction, the mild to moderately demented patient can be ambulatory with direct supervision until he is orthopedically safe to resume his usual level of function. When the patient is unable to follow instructions, particularly in regards to weight-bearing precautions, the wisest course may be to defer all rehabilitation efforts until sufficient stabilization at the fracture site is present to allow full weight-bearing. In this case, the physical therapist should encourage the nursing staff, family, or other care providers to try to maintain good positioning and prevent contractures until attempts at ambulation safely can be tried. Frequently, medications are given to the demented patient to decrease agitation, restlessness, or attempts to wander. The physical therapist should note whether these drugs impede motor performance and the rehabilitation process, and should notify the nurse and physician when this happens.

Katz et al. described the recovery of ambulation and activities of daily living (ADL) as a function of time. It must be noted, however, that this study was of a group of patients of above average socio-economic status admitted to a chronic disease, rehabilitation oriented hospital several weeks postfracture.[19] Walking was graded on a five point ordinal scale: (1) walking by self, (2) walking with mechanical aid, (3) walking with personal assistance, (4) walking with both mechanical and personal assistance, and (5) not walking at all. Activities of daily living were measured using the seven point ordinal Index of ADL which measures independence in bathing, dressing, toilet activities, transferring, continence, and feeding.[40,41] Of the 120 patients in this cohort[19] who had been independently ambulatory (without personal or mechanical assistance), 22 percent regained this ability by six months, 39 percent by one year, and 46 percent by 18 months. Only 2.5 percent regained independent ambulation in the period beyond 18 months. If patients requiring mechanical (but not personal) assistance are included in the group considered as recovering walking status, 68 percent had achieved this goal by six months, and only a further five percent of patients went on to reach this point by two years postfracture. Of the 15 patients who had required mechanical assistance in walking prior to their hip fracture, 40 percent had achieved their prefracture ambulatory level by six months, and 53 percent by one year. No further recoveries occurred in this group after one year.

Recovery in ADL occurred earlier than recovery in ambulation. Of 101 patients who had been totally independent in ADL prior to their fracture, 50 percent recovered this status in six months, and 55 percent by one year. When including patients who recovered to one or two points lower on the ordinal ADL scale as having an acceptable outcome, 70 percent had reached this goal by six months, and 72 percent by one year. Once having recovered function, it tended to be maintained. Recovered ambulatory function was sustained for 18 months or longer by 73 percent of patients, and recovered ADL function was maintained for at least this period by 72 percent.

HIP ARTHROPLASTY

Hip arthroplasty can be divided into three categories: (1) replacement of the acetabulum (cup arthroplasty); (2) replacement of the femoral head and neck (proximal femoral endoprosthesis or hemiarthroplasty); and (3) total hip-joint replacement (THR). The femoral endoprosthesis, both cemented and porous-coated, along with the THR will be discussed here in terms of surgical approach, recovery, rehabilitation, and postoperative complications. Emphasis will be on information of importance to the therapist for the successful treatment and rehabilitation with a hip arthroplasty patient.

Surgical Approach and Recovery

A detailed account of the surgical techniques for hip arthroplasty is not warranted here; however, a brief notation of the variations in surgical approach is appropriate. Surgical approach is quite often either anterolateral, transtrochanteric, posterolateral, or posterior. Much controversy among surgeons exists as to the best approach because each has certain advantages and disadvantages.[42,43] Even so it is important for the physical therapist to know which approach has been used in order to be aware of precautions against dislocations. It is also helpful to know whether a trochanteric osteotomy has been performed, as is often the case with the transtrochanteric approach.[43] Removal and rewiring of the greater trochanter make it necessary for the patient to observe certain postoperative precautions. (Specific dislocations and precautions for trochanteric osteotomies are discussed in the final section.)

Recovery of the hip arthroplasty patient with respect to postoperative pain levels, ambulatory skills, and return to previous activity levels is greatly affected by age of the patient, preoperative activity level, type of prosthesis used, pre-existing medical conditions, and postoperative complications.[44-46] No hard and fast rule can be used to predict recovery rates in hip patients, even though certain patient populations do not perform as well as others. As a result it is important for the therapist to consider all possible factors when treating the hip patient. A detailed patient history, knowledge of the surgery and postoperative complications, and a good rehabilitation program which provides for gradual increases in strength, range of motion, pain relief, and return to activities of daily living can do nothing but improve the patient's chances of return to presurgery level of function.

Rehabilitation for Hip Arthroplasty Patients

Exercises used in the rehabilitation of the hip patient are basically the same for all hip arthroplasties. With this in mind, the following program has been designed to encompass activities appropriate for all three groups. Modifications for special cases will be noted when necessary with the most obvious difference in treatment protocol among the three groups occurring in weight bearing status. The basic goals for rehabilitation following hip arthroplasty of any type should be the same:

1. relieve pain
2. maintain the integrity of the fixation
3. return to normal range of motion of the hip
4. return to normal strength in the involved extremity
5. achieve independence in ambulation
6. achieve independence in activities of daily living.

As mentioned previously, the degree to which these goals are attained will be affected by age of the patient, preoperative activity level, type of prosthesis, pre-existing medical conditions, and postoperative complications.[44-46] This in itself implies the need for an individualized program for each patient. Even so, certain guidelines can be followed in designing a postoperative regimen which can be modified for special cases.

Prior to beginning the actual rehabilitation program, the therapist generally sees the patient both preoperatively, in cases of elective surgery, and also in the acute postoperative setting. These are the times to begin preventive measures, i.e., instructions for deep breathing and coughing, lower extremity exercises, and positioning, which can help to minimize or decrease the magnitude of later complications. These activities should begin on the day of surgery and continue throughout hospitalization.[47] Lower extremity isometric exercises can include gluteal and quadriceps setting while isotonics can include ankle dorsiflexion/plantarflexion, ankle circles, or both. The deep breathing and coughing exercises help to maintain clear lungs and good abdominal muscle tone for trunk control. The lower extremity exercises coupled with elevation of both lower extremities are important in reducing edema and preventing venous pooling which can lead to thromboembolic disease.

Appropriate positioning of the involved lower extremity should begin early and be maintained throughout the first six postoperative weeks. In patients with posterior or posterolateral approaches, flexion past 60 to 70 degrees, adduction past the midline, and internal rotation should be avoided for prevention of dislocations. Proper positioning in bed can be achieved with an abduction pillow or regular bed pillows.[47] Elevated commode seats and pillows in chairs can aid in maintaining the correct degree of hip flexion in sitting. A pillow or cushion in a wheelchair with a hammock seat also can help to decrease internal rotation of the involved hip.

Bedside treatments should begin on the first day following surgery. Generally the patient may be allowed to sit up for brief periods (10–15 minutes) while completing ankle dorsiflexion/plantarflexion and ankle circles. Transfer training and gait training with a walker for short distances in the room are also appropriate at this time.

Depending on the preference of the surgeon, patients begin receiving therapy in the physical therapy department on the third or fourth postoperative day. At that time the patient can receive additional instruction in transfers and gait training with a walker and can begin to participate in mat activities. Appropriate positioning should be maintained during transfers, and active hip abduction of the involved extremity should be avoided in cases of trochanteric osteotomies. These patients should be trained to enter and leave the bed by leading with the uninvolved extremity in order to use hip adductors on the involved side.

Weight-bearing status at this time, for gait training either in the parallel bars or with a walker, will be determined by the type of prosthesis used. Generally a patient with a cemented prosthesis can bear full weight by day two or three as the cement is sufficiently strong at this point.[47] Even so, many physicians prefer to avoid full weight-bearing with a cemented hip for six to eight weeks. On the other hand a patient with a porous-coated implant will either be non-weight-bearing or toe-touch weight-bearing to allow for bony ingrowth in the implant since cement is not used. These patients will be maintained on toe-touch weight-bearing for one month with progression to full weight-bearing by the end of three months.[48] It is also important to note that patients with porous-coated implants generally require some type of assistive device in ambulation for a longer period of time than those with cemented implants. One surgeon suggests the use of at least one crutch for the first five months.[49]

Therapeutic exercises for the hip arthroplasty patient should include those which gradually will increase both range of motion and strength in the involved extremity. The exercise program should concentrate on all of the following areas:

1. hip flexion and extension
2. hip abduction and adduction
3. hip internal rotation (to neutral only)
4. knee flexion and extension.

The therapist should begin with active-assistive exercises for hip flexion/extension and abduction/adduction. Internal rotation to neutral should be done actively and knee extension with resistance should be completed if allowed by the physician. Based upon the patient's abilities, the amount of resistance and number of repetitions for each exercise should be increased gradually. Patients with trochanteric osteotomies specifically should avoid straight leg raising, bridging, and actively abducting the involved extremity for the first six weeks while the bone is healing. Some physicians prefer to avoid straight leg raising altogether.

The therapist should understand that hip flexor tightness can occur in the involved extremity as a result of positioning and extended periods of sitting. Activities to stretch the involved hip flexors should be an integral part of the program.

Finally, the patient should be given instruction on temporary modifications in life style that will be necessary in order to complete activities of daily living with as much independence as possible. The physical therapist in conjunction with the occupational therapist can train the patient in tub transfers, toileting, dressing, and other daily activities so that transition from hospital to home can be a smooth one. Progression of the patient to independence in activities of daily living along with a good home exercise program is essential for the successful rehabilitation of the hip arthroplasty patient.

Postsurgical Complications

In treating the patient following a hemiarthroplasty or THR, the therapist should be aware of certain signs and symptoms that may indicate postsurgical complications. Since some of the complications are potentially life-threatening, knowledge of common signs and symptoms coupled with appropriate treatment regimens is essential to the successful rehabilitation of the hip patient.

In many cases pain may be the only symptom present; thus, it is necessary for the therapist to outline the pain history including onset, site, aggravating conditions, and relieving factors. This information alone may be the major indicator of the pathology.[50] The most frequently encountered problems discussed here are infection, thromboembolic disease, nerve palsy, dislocation, heterotopic bone formation, fracture of the femur, and loosening of the prosthesis.

Infection of the hip following prosthetic surgery may lead to the need for revision surgery. Acute infection is often recognized early and can be treated with antibiotics along with incision and drainage with reasonable success.[51,52] Delayed infection is more difficult to diagnose. The patient will complain of pain of gradual onset in the area of the arthroplasty upon weight-bearing following a pain-free period of function.[50,51]

Thromboembolic disease occurs in an estimated 50 percent of

patients with hip fractures and arthroplasties.[46,51] When deep vein thrombosis (DVT) occurs in the lower extremity, the patient will complain of local pain in the area of the thrombosis. Edema often will be present along with redness and increased temperature of the extremity. These symptoms in conjunction with a positive Homan's sign (description found in reference 53) usually indicate a DVT. Patients suffering from DVT are treated with anticoagulant drug therapy. Preventive measures for DVT include postoperative elevation of the lower extremities, use of elastic stockings on the lower extremities, and early mobilization within the limits of the treatment program.[28,46,51]

Peripheral neuropathies following hip surgery are relatively uncommon but can cause problems in rehabilitation. Nerve damage can include the sciatic, femoral, obturator, or peroneal nerves[51,54] with subsequent loss of muscle function based upon the nerve involved. In most cases cited in the literature, muscle function either returned during postoperative hospitalization or within six months following surgery.[54]

Dislocation of the hip following surgery is the least frequent of the early postoperative complications.[55] Most often the dislocation occurs as the result of technical errors in the surgery itself, but other factors can contribute to the problem with the aged or confused patient being at high risk.[56] The direction of dislocation will vary based upon surgical approach and the type of prosthesis used. Anterior dislocations occur in a position of external rotation and extension of the hip. Posterior dislocations occur when the hip is internally rotated and fully flexed. A diagnosis can be made on the basis of acute pain, limited motion at the joint, and a characteristic rotational deformity based upon the direction of the dislocation. Dislocations can often be prevented simply by employing proper positioning. Patients with a posterior entry should avoid excessive flexion, adduction, and internal rotation of the hip in the first six weeks following surgery. Patients with an anterior entry should avoid excessive or forced extension of the hip with external rotation for the same length of time.[55]

Heterotopic bone formation has been found to occur in up to 40 percent of patients with THR. Even so, functional limitation is present in less than two percent of these cases.[46] Heterotopic bone formation is characterized by decreased range of motion at the hip either with or without pain.[57]

Fractures of the femur can occur upon insertion or relocation of the endoprosthesis.[52] Patients with this problem should be placed on a limited exercise and weight-bearing program until the fracture has fully healed.[58]

A late complication of hip surgery is loosening of the femoral component. The patient will complain of a gradual onset of pain with weight-bearing in the hip, thigh, groin, or knee. These symptoms may be accompanied by a "click" or "clunk" in the hip joint, and the patient will be unable to straight leg raise without pain.[52] The occurrence of loosening can be decreased by limiting the stresses applied to the fixation postoperatively. Limited weight-bearing while tissues heal, weight control by the patient, and practical limitations with respect to physical activity can all play a role in increasing the life expectancy of the prosthesis.[59]

CONCLUSIONS

Prior to the development of techniques for internal fixation of hip fracture, this injury all too often provoked the cry of the banshee. The rapid mobilization made possible through orthopedic stabilization has helped to diminish the development of complications of immobility. A significant number of patients, however, still either die or do not recover function, but most of these individuals previously suffered from additional disabling diseases. Life expectancy among those who survive the first few high risk months following fracture is considerable, and may be as high as six to eleven years.[9,19] Even slight variations in ambulatory and ADL function among the survivors can make a difference in the degree of dependency in a living situation. Living independently at home costs only a small fraction of requiring someone in the home daily or nursing home care. Multiplied by the years of life expectancy, it becomes clear that rehabilitation of these individuals is in the best economic interest of society, if indeed current rehabilitation techniques are effective.

This paper has presented physical therapy guidelines for caring for the patient with hip fracture. These guidelines are based on the opinions of clinicians frequently treating this condition.[15] While physical therapy is received by the vast majority of patients with hip fracture, we must rely on the opinions of experts in the field until we obtain the scientific evidence through research to guide our clinical decision-making. We need to learn for whom our costly service is beneficial and for whom it serves no need. What degree of physical therapy is needed by these patients, and for how long? How detailed should an evaluation be to identify the patient's problems, but avoid wasting time of the physical therapist? Is physical therapy needed just to get the patient up on a walker or should the patient continue to be followed in the transition to home and eventual free gait? Does gait training alone satisfy the needs of the patient or are range-of-motion exercises, strengthening, and ADL retraining also helpful? What outcomes should we look at? Should it be return to prefracture ambulatory level alone, or should we consider also the speed of recovery, the quality of gait, joint range of motion, strength, endurance, or independence in the broad spectrum of ADL? Can occupational therapy do a better job than physical therapy in assuring return to a safe and independent life at home once the patient is up on a walker? Will concerted efforts to involve the patient's social support network in his rehabilitation improve his outcome or the speed at which it is attained? Clinical trials which can control for coexisting determinants of outcome are best suited to answering these questions. The results of Katz's research[19] begin to suggest that rehabilitation is virtually completed by almost all patients with hip fracture at six months postfracture.

The second portion of this paper has presented physical therapy guidelines for the patient receiving either a hemiarthroplasty or a THR. These guidelines are based upon accepted protocol from physical therapists treating a great number of hip surgery patients along with input from orthopedic surgeons. It has been accepted readily in the medical profession that hip arthroplasty patients should receive physical therapy both before and after surgery with emphasis on progressive resistive exercises and gait training within the limitations of the individual patient.[46] Along with therapeutic exercise and gait training, the patient should also be trained in appropriate positioning techniques, activities of daily living, and modifications in activity level to permit return to an independent lifestyle with as few complications as possible. It is also important for the physical therapist to be aware of signs and symptoms of postoperative complications as some can be potentially life-threatening while others can lead to revision surgery. In essence, it is necessary for the therapist to consider the whole patient, including the past and present medical history, so that rehabilitation can begin before surgery and, if necessary, continue into the home setting.

REFERENCES

1. Mausner JS, Bahn AK: Screening in the detection of disease and maintenance of health. In: Epidemiology: An Introductory Text. Philadelphia, PA, W.B. Saunders Co., 1974, chap 11

2. Carter MC: A reliable sign of fractures of the hip or pelvis [letter]. N Engl J Med 305: 1220, 1981

3. Berger EY: More on diagnosing fractures of the hip or pelvis [letter]. N Engl J Med 306: 366, 1982

4. Alffram PA: An epidemiologic study of cervical and trochanteric fractures of the femur in an urban population. Acta Ortho Scand Suppl. No. 65:1964

5. Zimmer JG, Puskin D: An epidemiological model of the natural history of a disease within a multilevel care system. Int J Epid 4: 93–104, 1975

6. The old woman with a broken hip (editorial). Lancet ii: 419–420, 1982

7. Asher RAJ: The dangers of going to bed. Br Med J ii: 967–968, 1947

8. Bassey EJ, Fentem PH: Extent of deterioration in physical condition during postoperative bed rest and its reversal by rehabilitation. Br Med J iv: 194–196, 1974

9. Fitts WT, Lehr HB, Shor S, et al.: Life expectancy after fracture of the hip. Surg Gynecol Obstet 108: 7–12, 1959

10. Gordon PC: The probability of death following a fracture of the hip. Can Med Assoc J 105: 47–51, 1971

11. Beals RK: Survival following hip fracture: Long follow up of 607 patients. J Chronic Dis 25: 235–244, 1972

12. Sherk HH, Crouse FR, Probst C: The treatment of hip fractures in institutionalized patients. Orthop Clin North Am 5: 543–550, 1974

13. Pezczynski M: The fractured hip in hemiplegic patients. Geriatrics 12: 687–690, 1957

14. Thomas TG, Stevens RS: Social effects of fractures of the neck of the femur. Br Med J iii: 456–458, 1974

15. Hielema FJ: The Care and Cure of Hip Fracture in Nursing Homes. Ph.D. dissertation. Chapel Hill, North Carolina, University of North Carolina, 1981

16. Miller CW: Quality criteria for the treatment of hip fractures. Va Med Mon 102: 1032–1036, 1041–1043, 1975

17. Miller CW: Survival and ambulation following hip fractures. J Bone Joint Surg (Am) 60: 930–934, 1978

18. Kreutzfeld J, Haim M, Bach E: Hip fracture among the elderly in a mixed urban and rural population. Age Ageing 13: 111–119, 1984

19. Katz S, Heiple KG, Down TD, et al.: Long term course of 147 patients with fracture of the hip. Surg Gynecol Obstet 124: 1219–1230, 1967

20. Simon TL, Stengle JM: Antithrombotic practice in orthopedic surgery. Clin Orthop 102: 181–187, 1974

21. Morris GK, Mitchell JRA: Warfarin sodium in prevention of deep venous thrombosis and pulmonary embolism in patients with fractured neck of femur. Lancet ii: 869–872, 1976

22. Morris GK, Mitchell JRA: Preventing venous thromboembolism in elderly patients with hip fracture: Studies of low-dose heparin, dipyridamole, aspirin and flurbiprofen. Br Med J i: 535–537, 1977

23. Morris GK, Mitchell JRA: Can death from venous thromboembolism be prevented in elderly patients with hip fractures: (editorial). Am Heart J 95: 139–140, 1978

24. Sharnoff JG, Rosen RL, Sadler AH, et al.: Prevention of fatal pulmonary thromboembolism by heparin prophylaxis after surgery for hip fractures. J Bone Joint Surg (AM) 58: 913–918, 1976

25. Salzman EW, Harris WH, DeSanctis RW: Anticoagulation for prevention of thromboembolism following fractures of the hip. N Engl J Med 275: 122–129, 1966

26. Hollinshead WH: Textbook of Anatomy, ed 2. New York, Harper & Row, Publishers, 1967, chap 17

27. Gray H: Anatomy of the Human Body, ed 28. Philadelphia, Lea & Febiger, 1966, chap 5

28. Salter RB: Textbook of Disorders and Injuries of the Musculoskeletal System, ed 2. Baltimore, Williams & Wilkins, 1983, pp 89, 385–401, 539–546

29. Garden RS: Low angle fixation in fractures of the femoral neck. J Bone Joint Surg 43 (Br): 647–663, 1961

30. Pierce RO, Powell SG: The treatment of fractures of the hip by Roger Anderson well-leg traction, Clin Orthop 151: 165–168, 1980

31. Murray RC, Frew JFM: Trochanteric fractures of the femur. A plea for conservative treatment. J Bone Joint Surg 31 (Br): 204–219, 1949

32. Patrick JH: Intertrochanteric hip fracture treated by immediate mobilisation in a splint. Lancet i: 301–303, 1981

33. Hall G, Ainscrow DAP: Comparison of nail-plate fixation and Ender's nailing for intertrochanteric fractures. J Bone Joint Surg 63 (Br): 24–28, 1981

34. Wainer RA: Personal communication, 1984

35. Muckle DS (ed): Femoral Neck Fractures and Hip Joint Injuries, New York, John Wiley & Sons, 1977

36. U.S. Consumer Product Safety Commission: Home Safety Checklist for Older Consumers. Available from: U.S. Consumer Product Safety Commission, Washington, D.C. 202007

37. Hielema FJ: Hip fracture: An epidemic challenging the physical therapist. Phys Ther in Health Care I: pp. 49–58, 1986

38. Gucker T III: Exercise in Orthopedics. In Therapeutic Exercise, edited by Sidney Licht, Baltimore, Waverly Press, 1965, pp 641–671

39. Rusk HA: Rehabilitation Medicine. St. Louis, C.V. Mosby, 1971, pp 659–670

40. Katz S, Ford AB, Moskowitz RW, et al.: Studies of illness in the aged—the Index of ADL: A standardized measure of biological and psychosocial function JAMA 185: 914–919, 1963

41. Katz S, Downs TD, Cash HR, et al.: Progress in development of the Index of ADL. Gerontologist 10: 20–30, 1970

42. Roberts JM, Fu FH, McClain EJ, et al.: A comparison of the posterolateral and anterolateral approaches to total hip arthroplasty. Clin Orthop 187: 205–210, 1984

43. Vicar AJ, Coleman CR: A comparison of the anterolateral, transtrochanteric, and posterior surgical approaches in primary total hip arthroplasty. Clin Orthop 188: 152–159, 1984

44. Engh CA, Bobyn JD: Biological Fixation of the Porocoat AML Hip. Warsaw, Indiana, Du Puy, Inc., 1984, p 3

45. Long JW, Knight W: Bateman UPF prosthesis in fractures of the femoral neck, Clin Orthop 152: 198–201, 1980

46. Consensus Conference: Total hip-joint replacement in the United States. JAMA 248: pp 1817–1821, 1982

47. Charnley J: Low Friction Arthroplasty of the Hip: Theory and Practice. New York, Springer Verlag, 1979, pp 302–307

48. Engh CA, Bobyn JD, Matthews JGII: Biological fixation of a modified Moore prosthesis. In Hip Society: The Hip: Proceedings of the Twelfth Open Scientific Meeting of the Hip Society, St. Louis, The C.V. Mosby Company, 1984, pp 95–110

49. Harris WH: The porous total hip replacement system: surgical technique. In Harris WH (ed): Advanced Concepts in Total Hip Replacement. Thorofare, NJ, Slack Incorporated, 1985, pp 252–253

50. van Rens ThJG, Sloof TJJH: The investigation of the painful total hip. In Ling RSM (ed): Current Problems in Orthopedics: Complications of Total Hip Replacement. New York, Churchhill Livingstone, 1984, pp 231–241

51. Evarts CM, Gingras MB: Cemented versus noncemented endoprostheses. In Hip Society: The Hip: Proceedings of the Fifth Open Scientific Meeting of the Hip Society, St. Louis, The C.V. Mosby Company, 1977, pp 75–86

52. Gingras MB, Clarke J, Evarts CM: Prosthetic replacement in femoral neck fractures. Clin Orthop 152: 147–157, 1980

53. Hoppenfeld S: Physical Examination of the Spine and Extremities. New York, Appleton-Century-Crofts, 1976, pp 234

54. Ratliff AHC: Vascular and neurological complications. In Ling RSM (ed): Current Problems in Orthopedics: Complications of Total Hip Replacement. New York, Churchhill Livingstone, 1984, pp 23–24

55. Hamblen DL: Dislocation. In Ling RSM (ed): Current Problems in Orthopedics: Complications of Total Hip Replacement. New York, Churchhill Livingstone, 1984, pp 82–90

56. Sheppeard H, Cleak DK, Ward DJ, et al.: A review of early mortality and morbidity in elderly patients following total hip replacement. Arch Orthop Trauma Surg 97: 243, 1980

57. Concensus Conference: Total hip-joint replacement in Sweden. JAMA 248: pp 1822–1824, 1982

58. Barrington TW, Johansson JE, McBroom R: Fractures of the femur complicating total hip replacement. In Ling RSM (ed): Current Problems in Orthopedics: Complications of Total Hip Replacement. New York, Churchhill Livingstone, 1984, pp 30–40

59. Lee AJC, Ling RSM: Loosening. In Ling RSM (ed): Current Problems in Orthopaedics: Complications of Total Hip Replacement. New York, Churchhill, Livingstone, 1984, pp 134–145

Total Hip Replacement—
A Personal Perspective

Susan E. Roush., M.S., R.P.T

ABSTRACT. A physical therapist shares her personal experience of having had a total hip replacement. The history and course of degenerative hip disease is described as it led to the decision to participate in this orthopedic procedure. Insight into the decision making process is offered. The surgery and the author's response to postoperative treatment is described including perceived deficiencies in therapeutic management and communication skills. A complicated follow-up period provides the basis for suggestions for improving these deficiencies.

PHYSICAL THERAPY ORDER   Date: 7-24-80

22 year old female
Dx: Degenerative joint disease, left hip; admitted for total hip replacement.

Please see for preoperative instruction, postoperative follow up.

> ... seems like a routine order, I should be able to see her early this afternoon if I have a smooth morning. Let's see, hey she is pretty young for a total hip. I wonder what ...

It was a relatively routine order except for the fact that I was that 22 year old and the day before I was also a physical therapist who regularly picked up new orders.

HISTORY

Being 22 years old and having a degenerative joint problem almost seems contradictory. Except, of course, when the degener-

---

The author was an instructor in the Physical Therapy Program in the School of Health Related Professions, University of Mississippi Medical Center, Jackson, Mississppi. She is currently a doctoral student at the University of Washington.

ation occurs not as a result of prolonged stresses on normal structures but because of normal (or even less than normal) stresses applied to an abnormal foundation, as was true in my case. The underlying pathology associated with my degenerative disease was a congenital, dysplastic acetabulum that was not discovered until I was almost eight years old. The events leading up to the discovery were ordinary enough: I was learning to ride a bicycle and my parents noticed that every time I got off of the bike I was limping. I did not complain of pain, or remember falling, but the limp persisted, resulting in a visit to our family physician. X-ray investigation revealed the misshaped acetabulum which was evident to even the untrained eye. The head of the femur had started a slow migration toward the iliac crest, resulting in the visible limp.

Most orthopedic textbooks cite three to six months as the age when pelvic bones lose their plasticity so that reshaping (through casting or braces) is improbable or impossible. I only missed that by seven years. What followed for me after the diagnosis was heavy skeletal traction to pull the femur into a more normal alignment and an osteotomy procedure, known as a ''shelf,'' to stabilize the joint to accommodate weight bearing. The ''shelf'' was actually a bone graft from the iliac crest and required immobilization via a body cast, for five months. Cast removal brought physical therapy in the form of gait training, whirlpool, stretching and strengthening. The result was a sometimes painful but stable hip joint with limited motion; decreased external rotation and abduction were manifest functionally as a moderate gait deviation.

CHANGING SYMPTOMS

The stability of my hip held for about twelve years, at which time I had a year before obtaining my degree in physical therapy. My pain had been increasing in a quantitative more than a qualitative manner, and eventually led to a prescription for Motrin (an anti-inflammatory agent) and a cane for ambulatory support. A delicate balance had been reached that allowed me to continue most of my activities with minimal pain. It was a balance that would have to be repeatedly adjusted to accommodate fluctuating symptoms and the adverse side effects associated with treatment designed to relieve those symptoms, on one hand, and my social, vocational, and personal needs on the other. Minor changes on either side could upset this balance that was my "adjustment." I should not speak in past tense because it is a continual process, one I am sure most people experience from time to time; for persons with physical disabilities, however, the margin of error is considerably smaller.

The Motrin I was taking to control my pain ultimately led to gastritis, a 15 pound weight loss, and a trial of alternative medications. Other symptoms were also losing their subtlety: my endurance could rapidly be depleted on a simple shopping outing. Also, getting through an eight hour work day (I was employed as a staff physical therapist at a large acute care hospital by this time) was more difficult than I wanted to admit. A pattern had emerged in which I repeatedly sustained minor lumbar and midthoracic strains because I was not able to use my legs appropriately for lifting.

My "balance" was again out of kilter; the symptoms were gaining ground and the treatment had become more a part of the problem than the solution. Also, I continued to compromise my lifestyle to accommodate the pain. Medical, social, and vocational factors blended to form a situation I needed to change. I was lucky to know an orthopedic surgeon whom I trusted and with whom I could talk. This surgeon's advice, along with the input from an internist (who was monitoring my GI problems) and several physical therapists, coupled with my professional training and work experience, led to consideration of a total hip replacement.

TOTAL HIP REPLACEMENT?

My goals were threefold: (1) reduce the amount of medication I was taking and therefore diminish the adverse side effects; (2) decrease my reliance on a cane for ambulation; and (3) improve my gait and endurance. My surgeon saw several major risks associated with the surgery. The first was related to the fact that the cortex of my left femur was significantly atrophied. The left lower extremity had not received normal weight bearing since I was seven and Wolf's Law of bone metabolism easily predicted the result. The force required to implant the stem of the femoral component of the prosthesis securely into the shaft of the femur could fracture the femur itself. The second surgical risk also related to the anatomy of my proximal femur. The femoral neck was collapsed and the traction required to insert the prosthesis could produce a sciatic nerve stretch injury. A foot drop of unknown duration could be the result. Finally, infection is always considered a significant risk associated with any type of implant surgery. If not controlled, infection surrounding an artificial joint leads to removal of the components. Reinsertion may or may not be possible, and when it does occur produces results less positive than the original procedure. Another unsettling aspect of the risk of infection is that it requires lifelong monitoring and prophylactic measures.

Another risk associated with the surgery relates to the question of component longevity. These consequences are long term in nature, similar to those related to infection. Many questions surround the lifespan of artifical joints, and these unanswered questions make the procedures a controversial choice for "young" patients. I was certainly far short of the fifty or sixty year age bracket for which such procedures are routinely done. The life expectancy of the components (failure is related to loosening, breakage, and wear) is at best a guess. My surgeon estimated that mine could last roughly 10 to 20 years, but of course much less was possible (perhaps even probable) depending on a variety of factors including my activity level.

DECISION

I was in a situation where my life was being controlled by the symptoms, particularly the pain. Those symptoms were the focus of my life; I felt I had become a collection of malfunctioning body parts with a person along for the ride and not a person first who happened to be experiencing these symptoms. In essence, the pain had me, I did not have the pain. The surgery offered hope for control of the symptoms and consequently a more normal lifestyle. This control that I wanted was distinctly different from wanting symptom relief. (Of course I wanted to be pain free and throw the cane away forever but I knew that was impossible.) What I wanted was not to get rid of the pain, but just to be able to handle it. I wanted "me" back. Of course, a definite cost was associated with this possibility for improvement, in the form of making a lifelong commitment to unknown variables. The commitment was not just to this operation, at this time, but to making a "more normal" lifestyle compatible with the manmade replacement and questionable long-term consequences.

I made the decision to have surgery and plans were set in motion to make it a reality. I was granted a leave of absence from my job and would come back, pending surgeon approval, to administrative duties. I would gradually work back into patient care as tolerated.

SURGERY

My 12 day hospital stay was an experience in itself. Everyone has heard hospital "horror stories," but I was unprepared for a situation in which I was the target of insensitivity. But for all of the thoughtless comments and acts there were also helpful, competent professionals who were examples of health care at its best. The surgery itself, which involved a posterior-lateral approach, lasted more than four hours. The prosthesis, which had been custom designed to match my x-rays, was placed without sciatic nerve stretching or femoral fracture. Difficulty was encountered in lengthening the extremity to accommodate the new, normally shaped femoral neck. This lengthening caused the extremity to take on a position of excessive internal rotation. I never regained even the small amount of external rotation I had prior to surgery. My subsequent recovery and function would be hampered by the tightness of these internal rotator and adductor muscles.

My immediate postoperative period was very cloudy, filled primarily with sleep. I used a postoperative TENS unit to supplement pain medication. It was a device I was happy to have, not only for its pain reducing effects but because it gave me some control over my environment. This was particularly important to me because otherwise I felt relatively helpless in relationship to what was going on around me. This helplessness was coupled with a strong medication-induced haziness, that made me about as unresponsive to my environment as my environment was to me.

This lack of control and awareness was very much in evidence during the first several days of physical therapy. I had anxiously looked forward to "testing" my new hip after surgery but found myself to be a relative passive participant when physical therapy was started three days after surgery. I was aware of what was happening, but could not really help because I was in this fog; everything was done to me and only after five or six treatments did I become a more active participant in the therapy. And I knew the people who were working with me, knew the department, knew what was going to happen and why. Even with all of this

preparation, which typical patients rarely enjoy, I was quite unre-sponsive. When I did become more involved I was amazed at my lack of stamina and endurance: each treatment session would exhaust me. Progress seemed very slow, but by the time I was discharged I was walking independently with a crutch and was enjoying range of motion in my hip that I had not had since I was seven.

FOLLOW-UP

My recovery after discharge was bumpy. I was sent home with instructions to sleep and walk as much as I wanted and to use pain as an indicator of overactivity. I had thought my progress had been slow while I was hospitalized, but it felt as if it slowed to a snail's pace when I got home. Every gain seemed to be accompanied by two defeats. Again, my endurance and stamina proved inadequate to meet my expectations for daily living activities, and for returning to social and job activities. The soft tissue problems from my early postoperative days proved to be a significant limitation to my function because of pain. The deep joint pain of arthritis had been replaced with a more acute pain of muscle spasms. I was depressed, my surgeon and physical therapist were discouraged. There were no easy answers to the new problems, just as there had been no easy answers to the original problems. That precarious balance of symptoms and treatment on one hand and lifestyle on the other was tugged at, manipulated, and juggled until a workable compromise was reached. This balance took over a year to achieve and involved a less physically demanding career in physical therapy, traditional (e.g., drugs) and nontraditional (imagery) forms of pain control, and continued reliance on a cane.

PERCEPTIONS AND FEELINGS

I have had a unique opportunity to view the medical community both as a provider and as a consumer; during my experience as a patient I found out a great deal about its capabilities and limitations. The former overwhelmingly positive at times, the latter character-ized by an inability to look beyond its own ego. Let me try to illustrate several of my observations. One very frustrating occur-

84341

rence has to do with the system's apparent inability to deal with problems that are not directly from a textbook. Because I fitted into a particular box on a particular day, certain treatments would be prescribed without any recognition of the unique nature of my problem. If I would describe a certain symptom, for instance muscle weakness, corresponding treatment would be prescribed, strengthening. If that treatment did not work or produced side effects that outweighed the original problem, a treatment modification might be tried. This modification might take several forms but frequently results failed to meet expectations. When this happened the responsibility for the lack of results was directed back to me. The treatment approach certainly was not at fault, so it must be the patient's fault. This cookbook approach to health care does not work and it results in a demoralizing power play. And in any kind of power play that involves a patient and the medical system the winner is automatic. There are endless ways of communicating that your credibility has been lost; typically the words "learn to live with it" (translated "it is your problem not mine, stop bothering me") are passed out with a bottle of pills which seem to be an obligatory accompaniment. What happens when you are told "learn to live with it"? First of all, the symptoms are still there, usually worse because they have been the target of a lot of prodding and you have been told they will not go away. That is usually difficult enough to deal with, but you have also been slapped in the face with the suggestion there is something wrong with you or you would have gotten better. If you have been brought up in this society and have bought the image of the health care system as having all of the answers, it is easy to believe you deserved to be treated in such a manner. All because someone could not say "I don't know." Of course, there are people who do not get caught in these self-limiting emotions. I particularly remember a physical therapist I worked with after my surgery who did say "I don't know." His honesty reached beyond my pain and offered me understanding, hope, and acceptance. It was the attitude I needed to start putting the pieces back together. It was not his physical therapy expertise that helped me, it was his sensitivity and openness as a person. Didn't we all feel we could do that before they told us we had to know all of the answers?

The very system that is so reluctant to provide quality patient education to facilitate personal responsibility for wellness, jumps at the chance to pass off failure. Because that is what it is viewed as— failure. Problems are seen as having black and white solutions in a very gray world. It takes much more time, energy and creativity to truly work with a person to find those delicate balances that make disabilities minimally handicapping. But most health care practitioners are just not interested in that process.

Why does this happen? How has the system gotten so far off target? I feel there are many reasons health care providers misplace responsibility for lack of results. Certainly one reason is as a protection against appearing inadequate. If I present a medical problem that cannot be easily solved, that is a big threat to the "know all" image this society insists on holding for medical professionals. Never mind the complete unrealistic nature of that image.

I believe another reason associated with this practice is that unsolved medical problems are a direct reminder that we are all vulnerable to the abnormalities of the human body. Degrading terms such as "crippled" and "lame" typify the fear of vulnerability we all experience when we see a disabled person. "Thank God it is not me, it would be so awful." The labels keep that protective wall up that says it will never happen to me. It is very hard to get beyond that self-centered sentiment as long as the focus is on the issue of having a problem or not having a problem. Not having a hip problem is not an option for me, just as running the 100 yard dash in 9.5 seconds probably is not an option for you. The real issue is how can I be happy, healthy, and productive. Having a hip problem is just one of the many variables that makes up who I am and plays a part in how I seek those goals. Only when people can get beyond their fears can productive and meaningful relationships (professional or personal) develop.

These are my thoughts, today, on things that in one sense happened years ago and in another sense happen every day. I am sure my thoughts were very different when it was immediate, and I am equally sure they will continue to change as time passes. I hope my sharing has opened a small window of understanding between you and your patients.